COMMENTARY

CW00508087

on the

Third Edition of the

National Structural Steelwork Specification

for

Building Construction

Richard Stainsby DIC, CEng, FIStructE.

Published by: The British Constructional Steelwork Association Ltd,
4, Whitehall Court Westminster. London SW1A 2ES.

In association with:

The Steel Construction Institute, Silwood Park,
Ascot, Berks SL5 7QN.

THE BRITISH CONSTRUCTIONAL STEELWORK ASSOCIATION LTD

BCSA is the national organisation for the Constructional Steelwork Industry: its **Member** companies undertake the design, fabrication and erection of steelwork for all forms of construction in building and civil engineering. **Associate Members** are those principal companies involved in the purchase, design or supply of components, materials, services etc., related to the industry.

The principal objectives of the Association are to promote the use of structural steelwork; to assist specifiers and clients; to ensure that the capabilities and activities of the industry are widely understood and to provide members with professional services in technical, commercial, contractual and quality assurance matters.

The Association's aim is to influence the trading environment in which member companies have to operate, in order to improve their profitability.

A current list of members and a list of current publications and further membership details can be obtained from:
The British Constructional Steelwork Association Ltd,
4, Whitehall Court, Westminster, London SW1A 2ES.
Telephone (0171) 839 8566. Fax: (0171) 976 1634

 ## The Steel Construction Institute

The Steel Construction Institute develops and promotes the effective use of steel in construction. It is an independent membership based organisation.

Membership is open to all organisations and individuals that are concerned with the use of steel in construction, and members include designers, contractors, suppliers, fabricators, academics and government departments in the United Kingdom, elsewhere in Europe and in countries around the world. SCI is financed by subscriptions from its members, by revenue from research contracts and consultancy services and by the sale of publications.

SCI's work is initiated and guided through the involvement of its members on advisory groups and technical committees. A specialist advisory and consultancy service is available free to members on the use of steel in construction.

SCI's research and development activities cover many aspects of steel construction including multistorey construction, industrial buildings, use of steel in housing, development of design guidance on stainless steel and cold formed steel, behaviour of steel in fire, fire engineering, use of steel in barrage and tunnel schemes, bridge engineering, offshore engineering, and development of structural analysis systems.

Further information is given in the SCI prospectus available free on request from:
The Membership Secretary, The Steel Construction Institute, Silwood Park ,
Ascot, Berkshire SL5 7QN.
Telephone (01344) 23345. Fax: (01344) 22944

© The British Constructional Steelwork Association Ltd,
   4, Whitehall Court, Westminster, London SW1A 2ES.

Publication Number P 209/96
First Edition June 1996
ISBN 0 85073 029 5
British Library Cataloguing-in-Publication Data.
A catalogue record for this book is available from the British Library

Foreword

Many of us take pride in not conforming, we like to put our personal stamp upon what we produce; success often comes from our ability to find many different and diverse solutions to technical problems. However, within a contract, a degree of conformity is necessary even though this may be alien to our nature.

The National Structural Steelwork Specification for Building Construction provides documentation which can be included in a steelwork contract, and which helps to ensure that steelwork is accurately and economically made and can be safely built. It also lists the information which is needed such that a steelwork contract can be completed on time without recourse to contractual disputes.

This Commentary on the National Structural Steelwork Specification has been prepared to give guidance and information on the specification document and on the philosophy behind it. It does **not** form part of the specification, nor must it be used to interpret or construe any particular meaning to any of the clauses or sub-clauses in this specification.

Throughout this Commentary the National Structural Steelwork Specification for Building Construction is referred to as the NSSS.

Acknowledgements

A significant portion of this Commentary comes from notes made when compiling the original document. This volume therefore owes a great deal to the Steering Committee for the National Structural Steelwork Specification for Building Construction, chaired by Alan Watson.

Members of BCSA's Technical Committee have given helpful comments, as have the staff at BCSA. Peter Allen of BCSA also introduced me to the art and science of publishing. His ability to add humour to a critical situation has helped me to keep my feet on the ground.

Other friends in the structural steelwork industry have given useful advice. My sincere thanks are due to them all, but I do not burden them with any shortcomings in this Commentary; they are of course all mine.

Dick Stainsby

CONTENTS

Printed by The Chameleon Press Limited, 5-25 Burr Road, London SW18 4SG

INTRODUCTION

TERMINOLOGY WITHIN THE CONTRACT DOCUMENTATION

The National Structural Steelwork Specification for Building Construction (See Ref. 1) is intended to be a part of the contract documents alongside the Project Specification (which is described in Section 1) and other documentation dealing with commercial matters such as the Conditions of Contract. This can lead to a confusing clash of terms used in these documents. It is therefore essential to note the definition of terms given in the NSSS for the purposes of the technical specification, so that the meaning of each clause is fully understood.

The term **'Employer'** has a particular meaning in the NSSS and is so used to cover situations where the steelwork contract is a main contract, or a subcontract. Some 50% of all structural steelwork contracts for building structures are operated under a subcontract to a Joint Contracts Tribunal Standard Form of Building Contract (JCT), or a modification of it. The employer named in that document will be the person or company placing the main contract. However, as far as the NSSS is concerned the term **'Employer' is the individual or company placing the contract for the steelwork with the Steelwork Contractor**.

A similar problem can arise over the term **'Engineer'. The NSSS uses 'Engineer' to mean the person responsible for the structural design and for accepting the detail drawings and the erection method statement**. Other documents such as ICE Conditions or the FIDIC Conditions define the 'Engineer' as the person appointed by the Employer 'for the purposes of the contract'. These 'purposes' may include other duties not dealt with in the NSSS.

The reference to 'person' in the NSSS definition of Engineer is intended to be the designer as an individual, or a design company, or a design practice.

Design liability must be recognised when the Steelwork Contractor provides the design. In the terms of the NSSS, the Steelwork Contractor's designer has the responsibilities of the Engineer. This is an area where there is agreement between the NSSS and the JCT Standard Form of Contract With Contractor's Design 1981 Edition.

JCT deals with the matter under the heading 'Contractor's design warranty'. The essential part of the statement, in this respect, reads:

'the Contractor shall have in respect of any defect or insufficiency in such design the like liability to the Employer ..... as would ..... an appropriate professional designer holding himself out as competent to take on work for such design who acts independently under a separate contract with the Employer'.

'The Works' as defined in the Conditions of Contract are likely to include a brief description of everything which constitutes the owner's requirements for the completed building. **However in context of the NSSS, 'The Works' is limited to the structural steelwork.**

WHY HAVE A NATIONAL STRUCTURAL STEELWORK SPECIFICATION?

Over the years, those involved in drawing up contracts for the supply and erection of structural steelwork had felt the need to include a technical specification. British Standard specifications have been readily accepted by the industry for long enough, but what was needed, in addition to the national standards, was one document giving the requirements of the contract and which paid particular attention to those matters which are essential to good practice.

Some of these requirements had been considered too contentious for incorporation into British Standard Institution publications. Other matters of paramount importance in a steelwork contract could be found in a BSI Standard, but were lost amongst other information which it is necessary for a national standard to embrace. This resulted in most Employers composing a different technical specification each time. To become familiar with the technical requirements these bulky documents had to be read in detail whenever a new contract came along.

The concept of a national specification for structural steelwork arose in the 1980s, from discussions of a group which met under the auspices of the Department of the Environment and the National Economic Development Office, and the idea was readily endorsed by designers, specifiers and steelwork contractors throughout the industry.

BCSA undertook the development of such a specification under the guidance of a steering committee from all sides of industry. The objective was to achieve a greater uniformity in steelwork contract specifications by producing a document capable of being issued and cross referenced with both tender and contract documents.

The outcome of the work was the National Structural Steelwork Specification for Building Construction. The 1st. Edition was published in 1989 and the 3rd Edition, which this volume refers to, was published in 1994.

Having a nationally recognised document has proved that the wide fluctuation in contract specifications, which has previously been the cause of much misunderstanding, can be avoided.

PHILOSOPHY OF THE NSSS

A typical structural steelwork project starts with design. However, codes of practice for design are well established and such codes do not form part of the NSSS, or of other documentation in a contract. A reference to them is sufficient even when the contract is to 'design and build'.

Nevertheless, unless the design is well-ordered and thoughtfully made, it can play havoc with a steelwork contract and add considerably to the final cost. Due economy in materials used must be exercised, but it must be recognised that using a myriad of material sizes in order to save weight simply misses the economy which repetition and standardised details can bring.

The reason for introducing the subject of design in this statement on the philosophy of the NSSS is to make clear that although design is not within the scope of the document it is a factor which impinges on all sections of it. Successful steelwork fabrication and erection contracts depend upon the design being a practical one and also on the following items which are covered in the NSSS:

- information, supplied by the Engineer and the Employer, which is needed by the Steelwork Contractor so that he can perform - this is discussed under 'Project Specification' below

- the selection and proper use of good quality materials

- the preparation of detail drawings which are appropriate to the purpose of the structure and which take account of the need for tolerances

- the fabrication and assembly being carried out to a consistently high standard within recognised tolerances

- the surface of components being properly treated and handled

- the safe erection of the structure being carried out to an agreed programme, by a method which takes account of the design concept, within agreed tolerances

- the work being carried out to a quality plan.

True success calls for the work being performed economically. A perfect job which leaves either the Steelwork Contractor bankrupt or the Employer with excessive costs, will not foster a continuing structural steelwork industry. The NSSS is intended to be the key to the correct blend between what the Steelwork Contractor can economically achieve, and the price the Employer is willing to pay.

INSPECTION AND ACCEPTANCE CRITERIA

Included in the Welding Section of the NSSS are tables giving the scope of inspection and acceptance criteria for welds. These tables are the result of work commissioned by BCSA and carried out by The Welding Institute. Similar work has also been carried out for bridge structures under dynamic loads, but the tables in the NSSS are the first produced for building structures under static loading conditions. It appears likely that they will be incorporated into the European and International standards now being drafted.

EUROPEAN STANDARDS

The 3rd Edition of the NSSS makes reference to the increasing number of new European Standards which are replacing British Standards, a trend which will continue for some years to come. Further editions of the NSSS are likely to be needed simply to keep up to date with newly issued Standards.

New European (BS EN) Standards issued since July 1994, and Standards which are expected to be issued shortly, are listed on page 115.

The scope of the new European Standards generally does not completely coincide with the British Standards they replace, so some rearrangement of text and some rephrasing within the NSSS is expected to be needed from time to time. Some changes to the designated names of steel grades are given in the notes to the Materials Section.

A further matter which must be recognised is the adoption in European Standards of different words and phrases from those traditionally used in the industry, which are regarded as being more acceptable to those whose native language is not English. Some examples of terms used in design and specification are:

Traditional	European Equivalent
Dead Loads	Permanent Actions
Imposed Loads	Variable Actions
Unfactored Loads	Characteristic Actions
Capacity	Resistance
High Strength Friction Grip Bolts	Preloaded bolts
Fabrication and Erection	Execution

The British reader should be aware of the introduction of these terms and expect to see unfamiliar language appearing occasionally in new issues of documents as we move toward greater use of the European Standards.

THE PROJECT SPECIFICATION

Whilst it is intended that most general matters of a technical nature will be found in the NSSS, it is recognised that there must also be a place where particular requirements relating to the project can be brought to the notice of the Steelwork Contractor. This is the primary purpose of the Project Specification and it is the subject of Section 1 of the NSSS.

The Project Specification also provides a means of designating any changes or additions to the clauses in the NSSS (though it is hoped these will be few in number). When these changes are grouped together in the Project Specification they are unlikely to be overlooked *(See 1.1)*.

At an early stage of writing the NSSS, the decision was made that it would not cover specific surface treatments for structural steelwork. To have included the full range of paint coatings and other treatments would have resulted in an out-of-balance publication with a large section devoted to this matter alone. The answer was to include in the NSSS a short section on the rules for good practice in surface preparation, but to give the Project Specification the task of specifying the treatment to be used.

As the Project Specification is unique for each contract, it takes precedence over other specifications including the NSSS.

LEGISLATION

Although legislation does not form part of technical specifications such as the NSSS, it is nevertheless playing an ever increasing role in influencing the way in which manufacture and construction of a project should be carried out, and what type of materials should be avoided in use.

Whilst initially much of this legislation applied only to manufacturers and contractors, more and more is now beginning to apply to designers and the clients they work for. This commenced in the 1970s with the **"Health and Safety at Work Act 1974"** which laid down the duty of employers and employees to take reasonable care of the health and safety of themselves and others who may be affected by their actions or omissions at work.

Since the introduction of the 1974 Act in the U.K., other regulations, which arise from European Directives, are being converted into U.K. Regulations. Six of these statutes which set principles and general duties on employers and employees in all forms of working activities should be particularly noted.

They are:

- *Workplace (Health, Safety and Welfare) Regulations 1992*

- *The Management of Health and Safety at Work Regulations 1992*

- *Health and Safety (Display Screen Equipment) Regulations 1992*

- *Personal Protective Equipment at Work Regulations 1992*

- *Provision and Use of Work Equipment Regulations 1992*

- *Manual Handling Operations Regulations 1992*

On more specific aspects there is the **Control of Substances Hazardous to Health Regulations** which require the adequate control of carcinogens and similar dangerous substances. In parallel with this is **The Environmental Protection Act,** which sets limits to the discharge of those solvents, waste, etc. that can be harmful to the world about us. If the use of such materials can be avoided then costs may be reduced and risks to personnel having to handle them avoided.

The most recent piece of legislation emanating from a European Directive is the **Construction (Design &Management) Regulations.** These are outlined in more detail in Section 8 of this publication.

All British Standards carry a paragraph in their "Foreword" stating the position on legal liability when the Standard is used. The statement at present in use says: *Compliance with a British Standard does not of itself confer immunity from legal obligations.* This statement applies equally to the use of the NSSS.

QUALITY IN CONSTRUCTION

The aim throughout the NSSS is to set a high standard for quality in steelwork construction. This commitment to high standards and the maintenance thereof can be demonstrated by Steelwork Contractors having:

- a quality system demonstrating a company's management and production controls which, preferably, has been independently verified by, for example, *The Steel Construction Quality Assurance Scheme Ltd (SCQA).*

- an independent confirmation of a company's financial strength, resources and capability to undertake a particular type of work or size of project by, for example, registration on the *Register of Qualified Steelwork Contractors (RQSC).*

These two schemes, *SCQA* and *RQSC,* are complementary to each other and are not interchangeable.

QUALITY ASSURANCE

This is dealt with in Section 11 of the NSSS in which there is a requirement for all work to be carried out within a quality management system which is capable of ensuring that the Steelwork Contractor's responsibilities are executed according to the Contract and the Specification (NSSS).

Clause 11.1.3(ii) makes reference to BS 5750/EN 29000 Series Standards - it should be noted that since the publication of the NSSS 3rd. Edition these have been replaced by the BS EN ISO 9000: 1994 Series.

REGISTER OF QUALIFIED STEELWORK CONTRACTORS

This scheme, developed in conjunction with the Department of the Environment and which is supported by British Steel plc, sets competence criteria to be taken into account when assessing a company's ability to undertake steel construction projects. It is a system which is similar to schemes running in many countries worldwide. It is also intended to prepare the steel construction industry for linking to any Contractor Management Information System or Qualification Scheme which may be introduced for pre-qualification at a National or European level.

Companies joining will be included in the Register of Qualified Steelwork Contractors, which is available throughout the construction industry.

The Register's aims are:
* to ensure customer satisfaction;
* to enable clients to identify appropriate contractors for each project;
* to ensure that competing companies can be chosen within a set level of quality and value;
* to provide a system to establish contractors' compatibility with other European schemes and acceptability to other European countries.

There is the facility for companies to move to a higher class and category within the Scheme when they can fulfil the necessary criteria.

The two parts to the Scheme are:

Part 1 – Category   This refers to the type of steelwork for which a company is registered. Separate categories cover:

A   All forms of steelwork

B   Road/Rail bridges

C   Heavy industrial plant structures, including bunkers, hoppers, silos and supporting structures

D   High rise buildings (over 10 storeys)

E   Large span portals (over 30 metres)

F   Medium/small span portals (up to 30 metres) and medium/low rise buildings (up to 10 storeys)

G   Footbridges and sign gantries

H   Large span trusswork (over 20 metres)

J   Tubular steelwork where tubular construction forms a major part of the structure

K   Towers

L   Architectural steelwork

S   Small fabrications

Part 2 – Class   Companies are classified by the recommended maximum value of any single steelwork contract which can be undertaken. Registration also takes account of other financial criteria such as annual turnover, and experience in steelwork categories registered. Certification under a Quality Assurance scheme is indicated when applicable. A typical registration would appear thus:

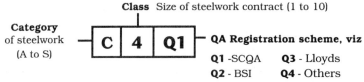

The Specifier wishing to appoint a Steelwork Contractor should designate a minimum registration 'level' for a particular steelwork contract and should request evidence from the prospective contractor that its registration meets this requirement. The documentation, which should be sent to the Steelwork Contractor for completion and returned to the Specifier, would typically be in the form illustrated on the next page.

When complete, please return this form to:

Specifier's Name .... Sir Joseph Bloggs and Partners

Contact Name .... A Bloggs C Eng

Address .... Construction House, 199 New Road,
NEWTON, Ampshire AB1 2CD

to be completed by the Specifier

Qualification as a Steelwork Contractor

In order to pre-qualify for steelwork projects, please supply the following information relating to your entry in the Register of Qualified Steelwork Contractors:

Classification

Class	Maximum Steelwork Contract Value
10	Up to £40,000
9	Up to £100,000
8	Up to £200,000
7	Up to £400,000
6	Up to £800,000
5	Up to £1,400,000
4	Up to £2,000,000
3	Up to £3,000,000
2	Up to £4,000,000
1	Above £4,000,000

Categories

Applicants may register in one or more categories to undertake the fabrication and the responsibility for any design and erection of:

A   All forms of steelwork
B   Road/Rail bridges
C   Heavy industrial plant structures
D   High rise buildings
E   Large span portals
F   Medium/small span portals and medium rise buildings
G   Footbridges and sign gantries
H   Large span trusswork
J   Major tubular steelwork
K   Towers
L   Architectural metalwork
S   Small fabrications

(a)   Category .... D and E

(b)   Classification 5

(c)   Date of last audit .... 16 - 3 - 1996

(d)   Copy of Registration Certificate Enclosed: Yes/~~No~~ *

(e)   Copy of BCSA Membership Certificate Enclosed (if applicable): Yes/~~No~~ *

(f)   Copy of QA Registration Certificate Enclosed (if applicable):   Yes/~~No~~ *

*   Delete as appropriate

Contact Name .... A N Other, Contracts Manager

Name of Steelwork Contractor .... Steelwork Structures Ltd.

Address .... Irongate Works, Blackthorne Industrial Estate
MIDDLEHAM EF3 4GH

to be completed by the Steelwork Contractor

Commentary on Section 1:
THE PROJECT SPECIFICATION

This section of the NSSS may be considered as the interface between other contract documentation and the NSSS. It describes information which is needed in order to undertake the work in fabricating and erecting steel structures. The Steelwork Contractor can only do the work and properly fulfil the contract conditions if the necessary technical information is complete and provided to an agreed programme.

The amount of information needed depends entirely upon the nature of the contract and the volume of work in it; at one extreme it may be a contract for the design, detail drawing, fabrication, protective treatment and erection of steelwork in a major project. At the other end of the scale it may be a simple contract to fabricate unpainted steelwork, designed by others and from drawings provided by others; all to be erected by someone else. Clearly much more information is needed for the former contract than for the latter.

Section 1 requires that the information is available in a purpose-made Project Specification which is produced specifically for each contract and that which forms part of the contract documentation.

Whatever the circumstances of the contract, the quality of the finished steelwork and the speed and smooth running of the project are dependent on the parties being properly informed. Assembly of the information listed will be a valuable insurance against argument, misunderstanding, and financial claims.

INFORMATION FOR THE STEELWORK CONTRACTOR

The information tabulated in Section 1 is that which is pertinent for most forms of contract used in the steelwork industry. Some of the items listed may be superfluous to some contracts whereas others, such as the subclauses dealing with the programme, will be part of every contract. The Employer should provide appropriate data to suit the project in hand.

Such information should not only be part of the contract documentation but, preferably, be part of the tender enquiry documents. Proper allowance can then be made in arriving at the tender price. If this is not done, the first area of a claim for extra costs in a contract can arise from differences between the assumptions made when preparing the tender and the information subsequently given on working drawings or in later documents.

As well as needing to know the quantity and complexity of the work, the Steelwork Contractor must be able to assess from the information given with the tender enquiry, the requirement for connection design, the degree of repetition, site conditions and the construction programme. An accurate tender price responding to these issues can then be prepared.

The Project Specification, being specifically written for the contract, takes precedence over all other technical specifications, including the NSSS.

1.1 NATIONAL STRUCTURAL STEELWORK SPECIFICATION

The intention is that the Project Specification should begin: *The National Steelwork Specification for Building Construction 3rd Edition, is included as part of this Project Specification.*

Any new or modified specification clauses can follow this statement; they can be given in the numerical form described below. In this way, the particular matters which are different to the NSSS can be made clear, preventing time consuming reading and searching.

Clauses which are to be modified should be covered in simple statements, retaining the numbering system with a suffix 'M' for modifications.

An example is:

The following clauses in the National Structural Steelwork Specification are amended as shown below :

3.2.2M   General Arrangement Drawings (Marking Plans)

Drawings shall be prepared by the Steelwork Contractor using CAD workstations. They shall show plans and elevations at a scale such that the erection marks for all members can be easily read when the drawing is plotted at A1 size.

3.7M   As Erected Drawings

On completion of the contract, the Steelwork Contractor shall provide the Employer with one set of paper prints of "As Erected" drawings comprising:

General Arrangement Drawings
Fabrication Drawings
Drawings showing revisions made after fabrication.

In addition a $3^1/_2$" floppy disk shall be provided of the
General Arrangement Drawings in DXF format.

New clauses should be introduced with a numbering system following consecutively after the Specification clauses, and with a suffix letter 'A' as in the following example:

"The additional clause given below is to be added to the National
Structural Steelwork Specification for the purposes of this contract:

5.7A   *Cruciform Butt Welded Areas*

Plates over 10mm thick to be incorporated in cruciform butt welds shall be
ultrasonically tested within 150mm of the weld, before and after welding
to ensure that they are free from laminations."

1.2   PROPOSED WORKS

If the Steelwork Contractor has a clear picture of the purpose of the building and is aware of its working environment, it will be easier for him to prepare appropriate design details and programme fabrication accordingly.

Steelwork which is to be used architecturally and which will be exposed to view should be drawn to the attention of the Steelwork Contractor. Other areas which deserve special attention by the Engineer, and which should be communicated to the Steelwork Contractor (as appropriate to his design responsibilities) are:

- where concealed steelwork in cavity walls may be exposed to moisture through condensation,

- where unusual loading pattens or intensities apply,

- where greater than normal deflection could be expected,

- where there is a possibility of settlement of foundations.

Defining the interface with other trades is also useful. The lack of this type of information often prevents the proper detail on drawings being prepared and could hold up fabrication whilst suitable amendments are made.

Where steelwork is measured in bills of quantities which are part of the contract documentation it is strongly recommended that the bills are cross-referenced to the steelwork design drawings.

(i)   A brief description of the structure

A concise description of the structural form of the building will enable the Steelwork Contractor to come to terms with the essential facts of the design. The most appropriate details and fabrication method can then be developed with less likelihood of missing some important detail. A description could be:

Five storey structure with braced end bays enclosed in cavity walls, with simply supported beams, shear studs welded to the top flange, carrying precast concrete units. The beams and floor units are designed compositely with insitu concrete placed to form "Tee-beam" construction.

or   *All columns have been sized by fire engineering design methods so that no fire protection is necessary. The inner faces of columns are exposed inside the finished structure.*

(ii)   The intended purpose of the structure

The finished structure will have a functional use which will sometimes be obvious from the title and drawings, but in other cases a description of the intended purpose may be necessary. This could simply be a statement such as:

The building is planned for retail shops with offices above.

or   *The structure is to house the production plant for the manufacture of animal foodstuffs, and there must be no exposed horizontal surfaces where dust can collect.*

(iii) Details of the Site in which the Works will be constructed

A block-plan giving details of site access for steelwork is a necessity. In addition any peculiarities of the site should be made known to the Steelwork Contractor. If, for example, the site is at the congested centre of a large chemical plant, rather than an easily accessible area with no restrictions, the Steelwork Contractor may have to change the procedure for erection.

(iv) A description of any temporary works required

There are occasions when the design is such that temporary works are necessary. An example is where floor beams have been designed compositely, but require temporary propping until concrete has set.

The site conditions may also dictate special temporary works being needed to facilitate erection. This could be, for example, a railway line which is to continue in use where temporary works to the structure must have sufficient clearance above and around the line. Matters of this nature must be communicated to the Steelwork Contractor.

Of course, the Steelwork Contractor also has a responsibility for the proper design of temporary works needed to suit his own erection scheme.

1.3   DESIGN AND DRAWING

The NSSS is not a design specification. However, a responsibility for design may be part of the steelwork contract. This section deals with the information required to undertake design and for making detail drawings. The level of responsibility is defined under three subtitles.

1.3.1   Design and detailing of the steelwork connections by the Steelwork Contractor after the member design has been prepared by the Engineer

The dividing line between member design and connection design has often been found to be a grey area in steelwork contracts. For this reason the NSSS gives a comprehensive list of information which is considered necessary for connection design to be made by a competent person.

The following notes amplify the list of items given in this clause but it must be remembered that the best results will be achieved when the connection designer works with the Engineer to produce the complete design of frame and connections. No matter how sophisticated the global analysis of the structure has been, the connections are crucial to its strength and stability; they could negate the design assumptions if the connection designer has not understood the Engineer's requirements.

(i)   **The statement** should make clear the parameters used for the design. The Engineer needs to communicate the following information to the Steelwork Contractor:

- the basis for the geometry of the frame,
- where eccentricity about grid lines has been introduced,
- where frame members have been positioned to achieve a particular clearance,
- where continuity is required,
- where rigid connections rather than simple connections are to be used,
- where a braced frame or wind-moment solution has been adopted,
- whether elastic or plastic design analysis was used.

The designer of the connections needs to be acquainted with all these facts if the design concept is to be met.

(ii)   **Design Drawings** are usually the best method of providing the Steelwork Contractor with information for connection design and are much preferred to design calculation sheets, computer output or long descriptive clauses.

A list of Design Drawings should be given in the Project Specification. Subsequent drawings made as the contract proceeds should be duly marked as being additional to those stated in the Project Specification, and the Steelwork Contractor advised which drawings have been updated and revised.

The degree of detail shown on the Design Drawings will of course depend upon the complexity of the structure, but the disposition of components in relation to each other, or in relation to essential architectural details, must always be given.

The Design Drawings must also give any special instructions for the safe erection of the structure perceived by the designer as being necessary due to the nature of the design. These will include the location of permanent or temporary bracing needed to safeguard the stability of the structure during erection, where the order and timing of erection may be important. Some components, such as trusses and deep plate girders, may be unstable until they are properly restrained, and beams in steel/concrete composite construction may be unstable until the concrete has cured.

It is good practice for the Engineer to provide on the Design Drawings typical details of the connections he envisaged when making his design. He should give particular attention to unusual or architecturally sensitive details. The Engineer is responsible for the design of the structure and this includes a responsibility for the type of connections used. The Steelwork Contractor is accountable for dimensional and arithmetic accuracy of his own work but the Engineer must ensure that the connections adopted are compatible with his design.

(iii)   Any **environmental conditions** which had to be taken into account in the design of structural frame will almost certainly apply equally to the design of connections, the selection of fasteners and the surface treatment to be used. The Engineer should communicate such conditions as are relevant to the Steelwork Contractor's responsibilities, and these might include:

- conditions which will be created by the processes which will take place in the building when it is used,
- greater than normal variation in temperature,
- local (micro-environmental) corrosion risks e.g. steelwork in cavity wall positions,
- erection access and transport problems.

(iv)   The **design standard** for the connections detailed must be that used for member frame design and this information should be given to the Steelwork Contractor. The Engineer may wish to introduce additional requirements of his own to be used for the design of connections. For example, he may call for lightly loaded members to have connections which will develop at least 15% of the member capacity in shear, or may specify that connections should comprise 2 bolts as a minimum.

(v)   **Restriction of methods of manufacture** are sometimes stipulated in design codes; for example, where plastic hinges occur in the structure, holes are required to be drilled and not punched. In cases such as this, the Steelwork Contractor, who does not have an intimate knowledge of the design analysis, must be advised of the location of these as they will mean a restriction on his manufacturing methods.

(vi)   The NSSS as a whole is intended for statically loaded structures. However, if specific components in a structure are to **resist vibrating or fatigue loading**, then the Project Specification must draw special attention to the design requirements for the connection of those components; a typical example would be a crane gantry which is to be used frequently as part of a production cycle, when full penetration welds and more critical weld acceptance criteria might be specified.

(vii)   The **forces, moments and their combination**, should be provided without the detailer having to study pages of design calculations or computer output. This will help to avoid errors in determination of the connection design parameters.

The Engineer should also ensure that loadings given for all load cases are compatible at node positions where several members meet. The connection design cannot be properly made if the forces in each member are not in equilibrium.

(viii)   When the design has been made using **limit state methods** it is usual to show reactions as factored forces on the Design Drawings. This should be stated in a note on the drawing to avoid uncertainty. *i.e Forces shown take account of appropriate load factors.*

See also the notes referring to foundation fixings and bolts in (xi).

(ix)   Figure 1.1 is an example where the member design should include consideration of the secondary stresses arising from connection design. Quite frequently the primary forces in the member must be known before the need for **stiffening elements** can be assessed. This is particularly true where there is an element of continuity at a connection. More often than not, the Steelwork Contractor will not know the primary forces in the member, so the Engineer should indicate any requirements for local strengthening.

Some indication of these requirements should also be made available at the time of tender since additional stiffening, due to circumstances not known when quoting, may lead to claims for recompense.

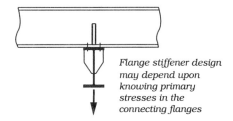

Flange stiffener design may depend upon knowing primary stresses in the connecting flanges

Figure 1.1 Flange stiffening

Unrestrained beams, which have to be notched where they connect to other members, may be particularly susceptible to instability. (See clause 4.3 in BS 5950.) It is therefore necessary for the designer of the member to state where notches will affect structural integrity (e.g. due to lateral torsional buckling) or where notches are specifically prohibited. An example can be found in *Joints in Simple Construction Volumes 1 & 2 (See Ref.2).*

(x)   **Grade 4.6 and 8.8 ordinary bolt assemblies** are both in common use. In the absence of a specific instruction, most Steelwork Contractors will use one or other at their own discretion.

The bolt finish may have to be compatible with the protective treatments used on the steelwork, and should therefore be specified in the Project Specification *(See 1.6 (v)).*

(xi)   Accurate design and detailing of **column base plates and holding down bolts** depends upon a knowledge of the foundation design. Generally the Steelwork Contractor is not given foundation design details and therefore cannot fully design connections to the concrete foundation.

If the Steelwork Contractor is required to design and detail the foundation fixings it should be so stated in the Project Specification. The necessary loads and overturning moments must be provided together with sufficient data about the foundations, e.g. grade of concrete, size of foundation block and position of reinforcement bars, such that holding down bolt positions, bond lengths and edge distances can be determined. In these instances it is necessary to state clearly whether such loads and moments are unfactored or factored values.

The steelwork Foundation Plan is prepared by whoever is undertaking the column base and holding down bolt design *(See 3.3)*.

(xii)   The connection design should conform to any specific restrictions imposed by the Engineer and, thereafter, should suit the economy of the Steelwork Contractor's own fabrication methods. Any restriction placed upon him should be made clear; since these will have cost implications, they should be made known at the time of tender.

As far as possible, the Engineer should allow the Steelwork Contractor discretion to use welded or bolted connections. This enables the Steelwork Contractor to plan the optimum use of plant and labour.

(xiii)   If **detailed requirements for cut-outs or appendages** required by other contractors only become available during the detail drawing stage, or even later, it should be recognized that there may be a delay in fabrication, and cost penalty for re-handling drawings and material.

It would be unrealistic to expect precise details at the tender stage; nevertheless, some indication of the details and quantities should be provided. Final details must be made available in time to suit the Steelwork Contractor's fabrication programme.

(xiv)   The essence of this requirement is that the deflection of the structure is determined by the Engineer. Therefore, any **camber or preset** to be provided during fabrication must be given in the Project Specification or on the Design Drawings, in order for the frame to achieve the required profile in service.

(xv)   **Punching holes** in steelwork is becoming more popular with modern numerically controlled machines in use. Areas where restrictions on punching apply for the reasons set out in 3.4.11 must be stated on the drawings. The specification for punching full size holes is in 4.6.3. This sub-clause is essentially covering the same ground as (v).

1.3.2   **Design and Drawing of the steelwork by the Steelwork Contractor commencing with sizing of members after the conceptual layout has been prepared.**

(i)   The **design requirements** which the Steelwork Contractor will need, when undertaking design in sizing members to a steelwork layout previously set by the Employer, may include: design concept (and basis for derivation of loads and moments), provision for mechanical and electrical services, preferred types of fire protection, provision for proprietary items, information regarding walls, roofs, ceiling heights and floors and fixing points etc. required on the steel framework.

(ii) **Drawings showing steel components** may be the architectural drawings, process plant drawings, or drawings indicating clear envelopes to be free of steelwork. Better results are achieved when there is close cooperation and regular communication between the Steelwork Contractor and the architect, or other party who has framed the structure.

(iii) The **design standard** to be used can be left to the discretion of a competent designer, however, some Employers may wish to state a preference. It is considered that at the date of this publication most designers will opt for using BS 5950, but some are still using BS 449 and the more adventurous have begun to use DD ENV 1993-1-1 'Eurocode 3: Design of Steel Structures'.

(iv) The **loading data** to be used may be a reference to BS 6399, but enough information on walls, partitions, roof and floors should be given to enable the Steelwork Contractor to properly assess the self weight of these items. The designer may be expected to use current British Standards for other loading such as snow and wind unless told otherwise.

(v) **Environmental conditions** which may affect detail design and drawing have already been described in notes to 1.3.1(iii) when discussing connection design

(vi) **Particulars of special details**, finishes or tolerances may be:

- where steelwork is to form an exposed feature - special care may be required in detailing, and this should be specified

- where tolerances more onerous than those set out in Sections 7 and 9 are required to accommodate cladding or plant, but it should be noted that it is often economic to consider the use of adjustable fixings so that normal tolerances can be applied to the structural frame.

(vii) **Material grade** for steel would normally be left to the Steelwork Contractor providing the design, but see notes to 2.1 for discussion on steel grades and their application.

(viii) Technical **specifications** for any non-steel items which are to be supplied by the Steelwork Contractor should be provided.

(ix) **Non-destructive testing** of material other than welds is not included in the NSSS but it is sometimes needed to ensure the integrity of the connection. An illustration of an additional clause of this kind, to be written into a Project Specification, is given earlier in this section (see commentary on 1.1).

(x) Different **deflection criteria** may be required to suit the fixing of cladding, or to control the lateral displacement of masonry walls.

1.3.3 Design and drawing of the steelwork by the Steelwork Contractor commencing with arranging the layout of members.

A great proportion of single storey building falls into this category. It includes retail shops and warehouses as well as structures for industrial processes such as those required for chemical plants and steel mills. When the Steelwork Contractor is providing design services in these areas, very close liaison with the Employer is essential.

In the case of industrial structures, steelwork costs are frequently a small fraction of the total cost of the plant and in these circumstances design of steelwork for economy may be overridden by more important factors affecting the operation of the plant.

(i)   **Conceptual drawings** may vary from simple sketches, showing no more than the building plan dimensions and height, to preliminary architectural drawings of a fairly detailed nature. If it is an industrial building the drawings may be expected to show the location of plant and the space available for the building structure.

(ii)   **Environmental conditions** which may have to be considered in preparing the layout and design of steelwork have already been described in notes to 1.3.1(iii) when discussing connection design.

(iii)   **Parameters** which may affect the design layout include:

- designing to accommodate the erection of plant within the structure,

- any requirements for clear spans, or limitations on construction depths,

- any requirements or restrictions on the location of bracings,

- the need to take account of the integrity of deep basements or retaining structures,

- difficult erection conditions necessitating limits to component size and weight.

(iv)   The **design standards** to be used can be left to the discretion of a competent designer as discussed in the notes to 1.3.2 (iii). However, when steelwork has to be designed to suit a specific requirement, the Engineer will need to specify which design standard is to be used for such structures or structural elements.

(v)   The **loading data** may be as described in the note to 1.3.2 (iv), or as required by the Employer and developed during the conceptual design stage.

1.4 WORKMANSHIP

The principal purpose of the NSSS is to set the normal standards required for good quality workmanship. The reason for this clause is simply to provide for special workmanship requirements which may sometimes be needed.

(i) The Steelwork Contractor must be informed if permanent **forms of identification**, like hard stamping, are not permitted for aesthetic reasons, or because they would be stress raisers if made at critical points.

(ii) Most welding undertaken by the Steelwork Contractor should normally be made to procedures approved by an independent authority in accordance with clause 5.3. However, the Engineer may wish to specify **special procedures** for welded details which have unusual features concerning material quality, thickness or geometry. He may want to approve sample welds made to the procedure before work on components proceeds.

(iii) **Fabrication or erection attachments** have to be properly drawn as required by 3.4.4. Welding of temporary attachments must also be of the same standard as for all permanent welds as stipulated in 5.4.5. In some cases, removal can be difficult and harmful to the structure and, in general, provided that such attachments are neat and unobtrusive, they should remain. Nevertheless, the Engineer may specify removal if he wishes.

(iv) Any requirement for **production test plates** to be tested by the Employer should be stated. When making butt welds, the Steelwork Contractor may use 'run-on run-off plates' as part of his welding technique. Sometimes these plates are made of a size suitable also for use as production test plates to prove the quality of the weld metal. These plates are not normally made available for testing by others unless this requirement is specified.

(v) The mandatory **scope of weld inspection** required by the NSSS is given in Table 1. It requires more inspection for welds which are used in connections than it does for longitudinal welds used in forming members assembled from plates and sections. If a greater degree of inspection than that given in Table 1 is required, it must be so stated in the Project Specification

Sometimes the design of welded connections may be more intricate than normal and therefore a call for 100% inspection of all such joints may be prudent; the Engineer must request this within the Project Specification.

Trial testing of welded shear studs is often needed particularly when the studs are site welded through the steel decking onto the beam below (See 5.6.2(i)).

(vi)   The **weld acceptance criteria**, given in Table 2, has been shown to be adequate for welds in structures not subject to fatigue or fluctuating loads. Wind loads may cause load reversal, but in practice the degree and frequency of fluctuation does not give rise to fatigue problems in orthodox structures. Different acceptance criteria may be specified for welded steelwork subjected to dynamic loads, such as crane structures.

1.5   ERECTION

Unless erection of a very simple structure is taking place on a greenfield site, the Employer may be expected to give comprehensive details about the erection site. Some examples are given below.

(i)   As well as indicating positions of datum levels, setting out points, access, working areas for cranes and storage areas, the **Site plan** should indicate other features which may affect erection. The Steelwork Contractor has the duty to arrange his work accordingly and to take account of them in preparing the Erection Method Statement.

These may include:

* proximity to areas of high fire risk,
* proximity to an airfield, or railway lines,
* proximity of overhead or underground services,
* peak traffic loads on adjacent roads,
* provision of artificial lighting,
* public access areas.

Reference should also be made to *Preparing the Site* in *Guidance Note GS 28 Part 2 (See Ref. 10)*.

(ii)   The Steelwork Contractor needs to know what **site services** are going to be made available for his use and where they are located. Details are also required of shared services, such as joint use of access or cranes, so that he can coordinate his activities with those of other contractors.

(iii)   A limit to the size of off-site assemblies may be necessary because of an access restriction at, or near, the site. A limit on weight may also be imposed on existing groundwork.

The length of vehicles used may be critical to the site access, and note should be made about impediments to access which may be caused by any low bridges on routes leading to the site.

(iv) If the **design features** are unusual or unconventional, any structural stability problems recognized by the Engineer should be communicated to the Steelwork Contractor. In such cases it will be necessary for the Engineer to describe the intended erection sequence he envisaged when preparing the design. See also notes to 1.5 (vii).

(v) Details of **underground services, overhead cables,** or any other **site obstructions** which may be encountered during erection, should be clearly shown on the Site Plan. It may also be appropriate to provide a 'clearance envelope' or to indicate measures being taken to protect underground services. If the site is contoured such that erection could be affected, this information should also be communicated appropriately on the Site Plan.

(vi) The Engineer should not presume that the Steelwork Contractor has any knowledge of the assumptions made in the design process, or of those regarding the behaviour of the structure.

Information should be provided when and where **special requirements** are needed for temporary bracing or propping. This could be a requirement to brace until concrete floors, which act as horizontal diaphragms, are in position, or until a stressed skin structure is clad. Columns in portal structures may require restraint at base level until adjustable foundation bolts have been grouted in position.

Temporary propping of beams may also be needed in composite steel/ concrete construction, or where they are supported by brickwork which has not attained its strength. Occasionally the wind forces can be greater on the bare frame, or partly clad frame, than on the completed building *(See Ref. 3)*. Care may have to be exercised by the Employer to prevent following trades stacking their materials indiscriminately on the framework.

The stage of construction when the frame is stable without temporary bracing should be indicated so that the Steelwork Contractor knows when the temporary bracing can be removed.

(vii) The advent of computer assisted methods of design analysis has enabled designers to create unusual and attractive structures which are highly indeterminate with several degrees of redundancy. However, the sequence of erecting components and the erection method envisaged is not always obvious; and **an outline** of the erection method must be provided by the Designer. Sub-clause (iv) deals with the necessity of ensuring stability of the structure during all stages of erection, but there may also be other precautions necessary to ensure that particular components are not over-stressed during erection.

Although the Steelwork Contractor is responsible for safe erection, he cannot be expected to know all about steelwork which is based on a new concept created by the Engineer. He is entitled to guidance on how the structure was expected to be built; this is an important obligation on the Designer under the C(D&M) Regulations.

Other constraints envisaged by the Engineer, such as phased erection programmes, or erection taking place on more than one front, must be specified.

(viii)   The **interface** between the Steelwork Contractor and other Site Contractors must be made clear by the Employer; programmes are particularly important. There must also be a clear definition of responsibilities in matters concerning safety; for example, where responsibility for the provision of scaffolding and handrailing lies, and provision of temporary access for each contractor.

(ix)   **Section 8** of the NSSS on erection includes safety on the erection site. Any other circumstances which may affect safety or site operations which may not be obvious to the Steelwork Contractor must be brought to his attention. It is not sufficient simply to call upon him to visit the site and expect him to seek out such matters that are relevant to safe erection. Note should also be taken of the C (D&M) Regulations discussed in the Commentary on Section 8.

1.6 PROTECTIVE TREATMENT

The NSSS does not give a menu of coating specifications from which a surface treatment can be chosen. The surface preparation required to precede any protective treatment must be given in the Project Specification. The coating system(s) should be specified so that the effect on the fabrication, transport and erection processes can be determined.

Guidance on various coatings including their life and maintenance characteristics are available in '*Steelwork Corrosion Protection Guides*' and in several publications from British Steel, BCSA, The Paint Manufacturers Association and the Zinc Development Association.

(i)   The options for **surface preparation** are usually:
- no preparation,
- wire brushing (manual cleaning),
- blast cleaning (either prior to or post fabrication, generally at the Steelwork Contractor's discretion).

(ii)   The options for **off-site or workshop protective treatments** are usually:
- no protective treatment,
- sprayed metal coatings (zinc and/or aluminium),
- hot dipped galvanizing (zinc),
- organic paint treatments by brush or spray (this can be priming only, or complete or partially complete coatings),
- duplex treatments with metallic and organic paint coatings.

(iii)   The options for **site protective treatments** are usually:
- no further treatment,
- hand touch-up of primers and applying finishing coats after erection,
- complete treatment system on site after erection including surface preparation; eg a combination of (i) and (ii) above.

(iv)   If the painting contract is split, clear definition is required of the **responsibility for cleaning down and repair** of any damage on site. It is usually more sensible and economic to have repairs carried out in the site painting contract but, in either case, the paint system must be 'repairable'.

(v)   The type of **protection for bolts** must be specified, whether by plating, galvanizing or sheradizing. If fasteners are protected with a film of oil this must be removed prior to painting.

(vi)   **Acceptance criteria** may consist of a statement of the permitted deviation in paint dry film thickness. A schedule of inspections at each stage of application can also be included. Some guidance on acceptance criteria is provided in BS 5493 *"Code of practice for protective coating of iron and steel structures against corrosion.'*

1.7   ADDITIONAL INSPECTIONS & TESTS

(i)   The Steelwork Contractor must be informed in the Project Specification if any of the tests required by the NSSS, or additional tests, are to be **witnessed** by the Employer, the Engineer or an Inspection Authority.

When the Employer intends to appoint an independent inspector, this requirement must be clearly defined because the timing and extent of testing can cause delay and increase costs *(See 11.2.)* .

(ii)   Neither party wants to waste time, so the **period of advance notice** required for these additional inspections or tests should also be stated.

1.8   PROGRAMME

(i)   The key to achieving completion of the erected frame in the specified time is undoubtedly the release of the construction issue of the design drawings at the agreed date or, for large projects, the phased release dates given by the Employer at the tender stage.

If further design development makes changes necessary to member sizes or dimensional adjustment, after the release of design information, claims for additional costs will arise.

(ii)   The Contract Documents should include a statement of the period for the acceptance/approval of the Fabrication Drawings. It should show a time in working days from receipt of drawings by the Engineer to comment being given. The time allocated should make due allowance for transmission time.

(iii)   The Steelwork Contractor needs to know the state of the site, when he can start checking foundations and when he can commence delivery to site. The site must be made ready to allow erection to commence on the due date, taking account of time required for any necessary remedial work to be carried out. This is necessary to avoid double handling material which has been fabricated to meet an agreed programme.

(iv)   Normally the date to start erection will have been agreed at tender acceptance stage between the Employer and Steelwork Contractor.

Where a phased programme is necessary, care should be taken to ensure that the time for each phase does not cause unplanned breaks in activity or periods of overlap.

In planning the steelwork erection programme the Steelwork Contactor will need to be told when following contractors are due to start and what requirements they have agreed with the Engineer in terms of completed steelwork.

Commentary on Section 2
MATERIALS

The essence of Section 2 is that materials must conform to a British Standard or, as is increasingly becoming the case, must accord with the European Standard which replaces the British Standard. However, reference must still be made to the British Standard where a European Standard has not been issued. The British Standard for a product may still be used until it is officially withdrawn.

The advent of European Steel Standards has resulted in an increase in the number of standards dealing with steel products. Whereas one standard previously dealt with all requirements of the product, it will now be found that steel qualities, dimensions and tolerances appear in different standards.

In the 3rd edition of the NSSS care has been taken to include the latest European or British Standard at the time of publication. However, with so many new standards replacing existing, a check should be made to establish changes since the NSSS was printed. In particular it should be noted that British Steel plc. now expects steel to be specified to the European Standards BS EN 10025, BS EN 10113 and BS EN 10155. *"New Steel Construction"*, published bi-monthly jointly by BCSA and SCI, lists details of new and revised Standards.

Cold rolled purlin sections and other cold rolled products are not specifically mentioned in Section 2. This is because they are usually supplied as proprietary items from manufacturers who specialise in their production. However, reference can be found in Table A in the NSSS to the steel material to be used for such components.

2.1   MATERIAL QUALITIES

Table A in the NSSS shows the appropriate British Standard or European Standard when the NSSS was last printed in June 1994.

It should be noted that the European Standards now designate steel grades by the design strength of the steel prefixed with 'S'; the designations design grade 43 and design grade 50 are now obsolete. Figure 2.1 shows the equivalent designation to be found in the European Standards and in former Standards.

The commonly used steels are grades S275 and S355 for which there are a number of sub-grades which refer to notch toughness qualities. Care should be taken when showing the required grade on the design drawings to ensure that the appropriate notch toughness sub-grade is shown.

BS 5950: Part 1	All BS EN's from 1993	BS EN 10 025 1990	BS 4360
Design Grade 43	S275	Fe 430	43
Design Grade 50	S355	Fe 510	50

Figure 2.1 Steel grades

BS 5950 states that the average temperature of internal steelwork in the UK may be assumed to vary from –5° to +35° (Celsius). When the service temperature for the structure is within these limits, much of the steel used may be expected to be either S275JR *(design grade 43B)* or S355JR *(design grade 50B)*. However, if the component is a tension member, it may require a Charpy impact value in line with one of the other steel sub-grades as a precaution against brittle fracture. Further guidance about choice of materials to avoid brittle fracture can be found in BS 5950 Part 1 clause 2.4.4 and Table 4 with cross-reference to the appropriate European Standards.

Sometimes it is necessary to design steel plates in higher strength steels than those provided by BS EN 10025. In these cases the appropriate standard for quenched and tempered steels (Q&T) is BS EN 10083.

The exceptions to the general rules about which design grade quality to use are:

- **Floor walkway plates in plain or Durbar pattern**. When properly stiffened or when designed to span nominal distances, walkway plates are not usually highly stressed as deflection is the controlling criteria. In these circumstances the Engineer may permit the use of a 'commercial quality steel' with a tensile strength (UTS) not less than 355 N/mm². BS 4592 gives further advice on all materials involved in industrial metal flooring, walkways and stairtreads.

- **Cold Formed Hollow Sections.** The calculation of structural capacity for cold formed material has a slightly different basis from that for hot formed material; for example, sectional dimensions and physical properties are different. When considering the use of cold formed hollow sections for primary structural members, it is recommended that appropriate care is taken its design and use *(see "New Steel Construction" June/July 1996, p. 30)*.

2.2   MATERIAL TESTING

The purpose of this clause is to draw attention to the fact that testing of steel by the manufacturer is a requirement of all steel material quality standards listed in Table A. The testing requirement includes a chemical analysis of the steel and determination of the yield stress and the tensile strength (UTS). Material elongation is also included as a percentage, and is a measure of the steel's ductility.

Dual test certification of steel to BS 4360 and European Standards was available from British suppliers until July 1995, but now all steel in common use is tested to the European Standards only.

2.3   TEST CERTIFICATES

The onus is placed on the Steelwork Contractor to ensure that all the steel he uses in the contract has a manufacturer's certificate. This includes steel which is obtained from a stock holder as well as that delivered directly from the mills.

The format of test certificates can be found in BS EN 10204 'Metallic Products - Types of Inspection Documents', and structural steel sections submitted to specific inspection should be type 3.1.B.

4.1.1 requires that all steel used must have a test certificate and is **traceable** to the certificate at the commencement of manufacture. However, there is no requirement for each finished element in the structure to be traceable to a particular certificate.

A typical inspection certificate type 3.1.B for a Universal Beam is illustrated in Figure 2.2. It should be noted that British Steel plc Certificates carry the following statement:

All Test Certificates issued by British Steel plc will contain either an embossed seal, or be impregnated with a British Steel plc watermarked logo, or a combination of both. Any recipient of a copy of a British Steel plc Test Certificate without the seal or watermarking should ensure from the supplier that it is a true and accurate reproduction of the original.

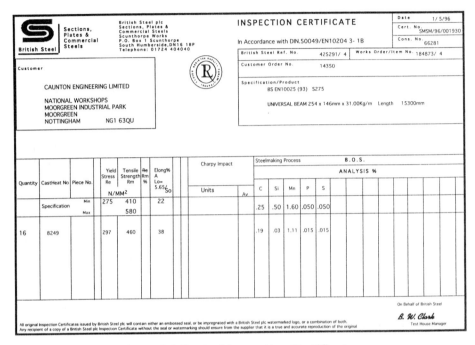

Figure 2.2 Typical Inspection Certificate

2.4   DIMENSIONS AND TOLERANCES

The dimensional accuracy of steel rolled sections is set by the steel manufacturer's rolling equipment and method of manufacture; it is not something which the Steelwork Contractor can change. The tolerances referred to in the NSSS give the permissible deviations of sections with respect to their shape, mass, length and straightness.

The tolerances with respect to the geometrical dimensions and mass are reproduced for convenience in Figure 2.3. It is only when preparing precise details for connections, architectural details etc. that the variations in Figure 2.3 need be taken into account. Generally, sections are well within their shape tolerance and this does not pose a difficulty. *(See also notes to sub-clause 3.4.6.)* It is not possible to have a section shape with every dimension at one extreme of the limits given in Figure 2.3, as it would fall outside the limitation on mass.

An under-rolling with respect to thickness is generally offset by the yield

Tolerances for Universal Beam and Column Sections
to BS4: Part 1 which refers to BS EN 10034

Section height h (mm)	+	−	Width b (mm)	+	−	Web thickness s (mm)	+	−	Flange thickness t (mm)	+	−	Mass
h ≤ 180	3.0	2.0	b ≤ 180	4.0	1.0	s ≤ 7	0.7	0.7	t ≤ 6.5	1.5	0.5	
180 < h ≤ 400	4.0	2.0	110 < b ≤ 210	4.0	2.0	7 < s ≤ 10	1.0	1.0	6.5 < t ≤ 10	2.0	1.0	
400 < h ≤ 700	5.0	3.0	210 < b ≤ 325	4.0	4.0	10 < s ≤ 20	1.5	1.5	10 < t ≤ 20	2.5	1.5	
h > 700	5.0	5.0	b > 325	6.0	5.0	20 < s ≤ 40	2.0	2.0	20 < t ≤ 30	2.5	2.0	± 4%
						40 < s ≤ 60	2.5	2.5	30 < t ≤ 40	2.5	2.5	
						s > 60	3.0	3.0	40 < t ≤ 60	3.0	3.0	
									t > 60	4.0	4.0	

Tolerances for Joist and Channel Sections
to BS4: Part 1

Section height h (mm)	+	−	Width	Web thickness	Flange thickness	Mass
h ≤ 3050	3.2	0.8	none given	none given	none given	± 2.5% or
305 < h ≤ 381	4.0	1.6				+ 5% when
381 < h ≤ 432	4.8	1.6				min. mass ordered

Tolerances for Equal and unequal angles
to BS4848: Part 4 which refers to BS EN 10056-2

Leg length a or b (mm)	+	−	Thickness t (mm)	+	−	Mass
a ≤ 50	1.0	1.0	t ≤ 5	0.50	0.50	± 6% for
50 < a ≤ 100	2.0	2.0	5 < t ≤ 10	0.75	0.75	t ≤ 4 mm
100 < h ≤ 150	3.0	3.0	10 < t ≤ 15	1.00	1.00	
150 < h ≤ 200	4.0	4.0	t > 15	1.20	1.20	± 4% for
a > 200	6.0	4.0				t > 4 mm

Tolerances for Circular Hollow Sections
to BS4848: Part 2

Outside Diameter D(mm)		Mass
D > 50	± 0.50 mm	±6%
D > 50	O.1D	±6%

Tolerances for Rectangular Hollow Sections
to BS4848: Part 2

Outside Dimensions a or b (mm)		Mass
a > 50	± 0.50 mm	±6%
a > 50	O.1D	±6%

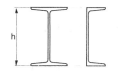

Figure 2.3 Section tolerances

strength being higher than the guaranteed minimum. Thus, the section properties, which are based on the sectional dimensions in the British Standard and given in the various publications from BCSA, SCI and British Steel plc, are suitable for design purposes.

2.5 SURFACE CONDITION

This clause is mainly designed to ensure that steel used in a contract is not heavily rusted or pitted. It refers to BS 7079 Part A1 which requires a visual assessment of surface condition to be made prior to fabrication. This British Standard shows representative photographs of surface appearance after various cleaning methods; and **2.5(i)** of the NSSS stipulates Rust Grade C as the minimum surface quality.

Parts **(ii)** and **(iii)** of this clause deal with other surface imperfections which may be revealed during surface cleaning; rectification is also covered in the Standards. It will be found that, in some cases, surface grinding will suffice, whereas in others, welding followed by grinding will be necessary.

2.6 SUBSTITUTION OF MATERIAL OR FORM

From time to time it may be prudent or necessary to change materials from those specified. There may be good reasons for doing this such as availability of materials, or so that certain section sizes can be ordered in larger quantities to avoid the penalty of higher costs per unit for small quantities. It is demonstrable that the process of rationalising the range of sections used in a project can lead to considerable benefits.

Provision is made here for such substitutions, but they may only be made with the agreement of the Engineer.

The Engineer must decide whether to allow a substitution after considering all the design parameters and programme requirements as well as any legal and contractual aspects.

It should be noted that hot-rolled and cold-formed hollow sections are seldom directly interchangeable. If the design shows 'hollow sections' then it must be assumed that 'hot formed hollow sections' are required unless further enquiries are made *(see comments to 2.1 for further notes)*.

2.7   WELDING CONSUMABLES

The use of correct welding materials is essential to the achievement of satisfactory welds. **2.7.1** advises the **British Standards** for consumables used for manual metal arc, gas shielded arc, and submerged arc welding processes, while **2.7.2** gives the requirements for **storage** of welding consumable materials.

Covered electrodes are susceptible to moisture which can then be the cause of hydrogen cracking if used in a damp state. It is, therefore, essential to store them in dry and warm conditions as specified in BS 5135. The issue of welding electrodes to welding operators should be controlled to ensure that the correct electrode is used and that they have not become damp before use.

2.8   STRUCTURAL FASTENERS

Details of the appropriate British Standards applicable to bolt assemblies used in the structural steelwork industry are given in **2.8.1** to **2.8.6**.

In **2.8.7** the responsibility for the supply of fasteners with specified **coatings** is placed on the manufacturer. The purpose of this is to ensure that the manufacturer safeguards the product against any applications of heat when applying the coatings which might affect the steel properties. BS 7371 covers the various protective coatings which may be used on metal fasteners *(see 1.6(v) and 2.10)*.

Fully threaded bolts, technically referred to as screws, are not specifically dealt with in the NSSS, although their use is encouraged. A rationalised range of fully threaded bolts is recommended as follows:

Grade	Diameter	Lengths (mm)		
*4.6	M12	25	-	-
8.8	M16	30	45	-
8.8	M20	45	60**	-
8.8	M24	70	85	100

*   *Intended for use in fixing cold rolled purlins and rails*

**   *It is estimated that 90% of bolts used in a typical structural frame can be 8.8 M20 x 60 fully threaded bolts.*

Bolts and screws should now be specified to one of the European Standards:

Grade	Bolts	Screws	Nuts
8.8	BS EN 24014	BS EN 24017	BS EN 24032
4.6	BS EN 24016	BS EN 24018	BS EN 24034

(Fasteners are still available to BS 3692 and BS 4190 although these Standards have been declared obsolete.)

The normal washer, when required, for structural steelwork is 'Form E' as described in BS 4320 'Specification for metal washers for general engineering purposes'. These are 'Normal' washers, as opposed to 'Large' or 'Extra Large' washers which are also included in the Standard.

Washers suitable for use with holding down bolts are 'Extra Large' type (Form G). They are thicker than other washers, and have an outside diameter which is approximately three times the bolt diameter.

2.9   SHEAR STUDS

There is at present no European or British Standard for **Shear Studs** except for those used in bridges. The NSSS therefore gives the mechanical requirements for studs when used in building structures.

2.10 PROTECTIVE TREATMENT MATERIALS

Details of the appropriate standard for blast cleaning materials, sherardized coating and zinc for galvanizing are given, but the standard for paint treatments and other coatings are not stated.

The intention is that all coating materials used in protective treatment systems are in accordance with the recognised Standard for the material. Standards commonly referred to are:

BS 729	Specification for hot dip galvanized coatings for iron and steel articles
BS 3698	Specification for calcium plumbate priming paints
BS 4147	Specification for bitumen based hot applied coating materials for protecting iron and steel, including suitable primers where required.
BS 4652	Specification for zinc-rich priming paint (organic media)
BS 4800	Schedule of paint colours for building purposes
BS 4921	Specification for sherardized coatings on iron or steel
BS EN 22063	Metallic and other inorganic coatings. Thermal spraying. Zinc, aluminium and their alloys

2.11 PROPRIETARY ITEMS

It is common for many different proprietary items to be specified by the Employer and supplied as part of a steelwork contract. These may include types of roofing materials, cladding and glazing, as well as specialist coating materials for surface protection or fire protection.

The NSSS simply requires the Steelwork Contractor to ensure that all proprietary items are used in the manner prescribed by the manufacturer.

If the Engineer specifies proprietary items on behalf of the Employer, it is his responsibility to ensure that they are appropriate for their intended use.

Commentary on Section 3
DRAWINGS

Section 3 is intended to apply to all drawings made as part of the Contract by the Steelwork Contractor. It aims to ensure that they are clear and legible and made to a recognised format.

Although produced primarily for fabrication and erection, the drawings must also be clear and unambiguous to the Engineer. However, the Steelwork Contractor is not obliged to add additional information to the drawing to assist the Engineer. In other words the drawing need only show what is necessary for steelwork fabrication and erection and for the Steelwork Contractor's own checking procedures (see *the provisions in 3.2.2*).

The drawings have a critical role to play in the execution of the contract because they are still the principal way of transferring information and it is a golden rule that the Steelwork Contractor's workforce does not depart from what is shown on the drawings.

Design drawings made by the Engineer feature in 1.3.1(ii), and are equally important to the success of a steelwork contract. The Steelwork Contractor is entitled to expect that such drawings are made to a similar standard and quality as the fabrication and erection drawings.

3.1   GENERAL

BS 1192 "Construction drawing practice" contains recommendations for the preparation of all drawings for the construction industry including structural steelwork. **3.3.1** therefore specifies it as the Standard to be used. Part 1 provides recommendations for general principles, Part 2 gives recommendations for architectural and engineering drawings, and Part 3 the symbols etc. which are be used.

Welding symbols are not illustrated in BS 1192 Part 3 so it cross refers to BS 499 Part 2. However, BS 499 has now been largely replaced by BS EN 24063 and BS EN 22553.

Part 5 of BS 1192 deals with computer drawings and the exchange of data between CAD users, but this method of CAD exchange, although growing in use, has yet to gain wide acceptance, not due to technical data transfer problems, but for reasons such as liability for accuracy of the data.

More comprehensive information on making drawings can be found in *The Steel Detailers' Manual (See Ref. 4).*

3.1.2 requires that when **revisions** are made to drawings, that a new suffix letter is added to the drawing number. Details of the revision must also be adequately described in the notes on the drawing. It is **not** sufficient to state "revised", or "general revisions". A brief, but specific description of changes made should be shown in a section of the drawing devoted to revisions.

Although not specified in the NSSS, it is good practice to provide a 'cloud' around a revision so that it will be easily seen when the drawing is being read. For example, a revised dimension may be shown as in Figure 3.1.

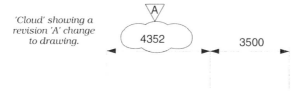

'Cloud' showing a revision 'A' change to drawing.

Figure 3.1   System for indicating revisions

It is stipulated in the NSSS that the drawing office system must cover the issue of revised drawings and withdrawal of the previous copies.

3.2   GENERAL ARRANGEMENT DRAWINGS

The General Arrangement drawings are used essentially as marking plans and are therefore the key drawings for the erection of the structure. For this reason the NSSS places an emphasis on the requirement that all members have a clear marking system which ensures that they are easy to locate.

A typical General Arrangement drawing with unique member references is shown in Figure 3.2. The position of the mark, on the face of a column and end of the beam, determines the physical direction of the member when erected. In the example illustrated the column marks are on the East face and the beam marks are consistently at the North end, or East end, of the member. This facilitates easier erection since the erector can then orientate the member without having to examine the drawings.

Columns may be identified with an 'S' mark, enabling identical columns to have the same mark, or by using the plan grid reference coordinate as the stanchion mark for that member. (Both systems are shown in Figure 3.2.)

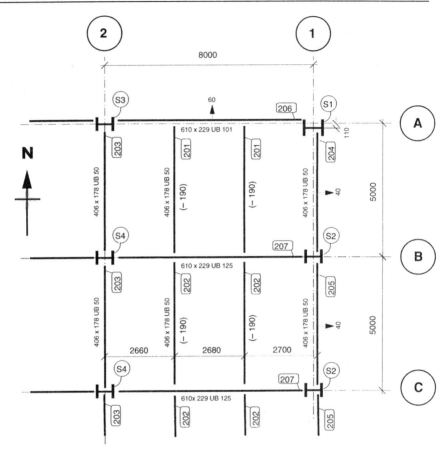

Figure 3.2   Part of a typical General Arrangement Floor Plan

Dimensions should be given on the drawing to all grid lines with any eccentricities indicated. Beam levels should be plainly shown. This may be by reference to a finished floor level with a distance above or below it given, or another clearly understood system. Raking and sloping members should also be properly located with principal dimensions. Enlarged details should be provided where necessary to ensure proper location.

For large structures when the G.A. drawing comprises several sheets, it is expedient to show a key plan of the part-drawing in relation to the whole. It is also useful to include on the G.A. a note of the last mark used for each type of component, so that new component marks can be readily added as detail drawing proceeds.

3.3   FOUNDATION PLAN DRAWINGS

The responsibilities for the design and detailing of the steel base plate and holding down bolt connections to the concrete foundations are discussed in 1.3.1(xi) of this commentary. Responsibility for the preparation of the Foundation Plan will depend upon the agreement reached, but it must contain all the information required by this clause.

Bolts arranged in sleeve pockets for adjustment must be detailed such that the bolt adjustment on plan (within the tolerance give in 9.4.3) can be achieved before final grouting.

3.4   FABRICATION DRAWINGS

The intention of **3.4.1** is to make clear that **Fabrication Shop Drawings** are primarily intended for the manufacture of components and should not include unnecessary information. Data necessary to simplify checking, such as grid line dimensions, may be shown but other details irrelevant to the checking or manufacturing process should be omitted.

It is normally expected that Fabrication Drawings will show components as completed for dispatch from the fabrication shop.

Components and fittings produced on automated plant are usually made directly from electronic information data. In this case Fabrication Drawings are used for shop assembly only.

Although not stated, it is expected that all hand made drawings are checked by a second competent draughtsman and the initials of the draughtsman and checker must appear on the drawing. However, in the case of drawings generated by computer, it is considered to be more important to check the computer input information plus a visual check of the drawings produced to ensure that no obvious errors have been made.

3.4.2 is designed to ensure that the management of manufacturing drawings is properly controlled through a **drawing register**. Information contained in the register will include the dates when the drawing is sent for approval and the issuing dates for each revision.

The NSSS requires an index system to show the components to be found on the drawing, by listing the erection marks. However, it is more common nowadays to have single component drawings with the drawing number being the same as the erection mark.

The advent of computer drawing is making **microfilming** systems for drawings based on 35 mm negatives obsolete. Nevertheless, **3.4.3** gives the reference to BS 5536 'Recommendations for preparation of technical drawings for microfilming.'

3.4.4 is included in the NSSS to ensure that **attachments** which have to be provided to facilitate erection are shown on the Fabrication Drawings, so that such attachments can be properly positioned and checked.

It may also be necessary to locate the fitting on the centroid of the component, and any such requirements should be clearly specified on the drawings. The Steelwork Contractor should include a reference to erection attachments in the erection method statement, since the Engineer must give his approval to all design and stability matters in the statement. The requirements for the erection method statement can be found in 8.1.1.

The intention of **3.4.5** is to reflect the importance of **preparations for welding** as part of the connection by having them shown on the drawings. However, if a Steelwork Contractor uses a particular preparation regularly, he can reference this on the drawing without showing the detail. There is no requirement to show weld procedures on the drawings.

The fabrication workshop's procedure is expected to include weld inspection as specified in 5.5.5, but if a different level of inspection has to be provided, the NSSS stipulates that such information be given on the drawings.

The art and science of good detailing practice for fabrication drawings is, in many ways, encompassed in **3.4.6**. The NSSS expects that the draughtsman has a full knowledge of the permitted tolerances. His job includes showing **packings and clearances** required so that the component can be erected without modifications having to be made at site.

If the geometry of the completed structure is to be as shown on the Design Drawings, detailing may have to take account of camber or flexure which will occur from the self weight of the structure. It is up to the Engineer to state if this is needed; it is not difficult for the detailer to make provision when the camber requirements are known. It is sometimes necessary to make provision for adjustment in other connections, for example:

- vertical cantilevering parapet posts connecting to portal frame columns.

- brackets supporting gutters requiring adjustment to accommodate a drainage fall.

The clearance requirements for **hole sizes** in **3.4.7** are essentially those commonly used throughout the industry. The 6 mm hole clearance for holding down bolts allows for normal adjustment. However, in thick steel base plates, or when the bolt extends into a bolt-box welded to the column shaft, an even greater clearance may be needed where positional adjustment may incline the holding down bolt.

6.1.8 stipulates a clearance of only 0.3 mm for precision fitted bolts. It is expected that these will be used in holes which are drilled after assembly, or pre-drilled undersize and reamed when assembled.

The NSSS does not mention short horizontal slot in cleats as shown in Figure 3.3. Some Steelwork Contractors may wish to adopt this detail which improves the adaptability of cleats and allows a standard cross centre dimension where web thicknesses vary. This detail can be used only when accepted by the Engineer.

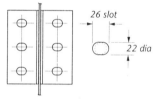

Figure 3.3 Short slot holes

Connections where adjustment is required, or where movement is necessary, need special attention when preparing detail drawings. Two such matters are the subject of **3.4.8 and 3.4.9**. Holding Down Bolts, requiring greater facility for adjustment than other bolts, have to be provided with loose **cover plates** or large washers *(See note to 2.8.5)*.

It is important that a bolt assembly in a **movement connection**, such as used at expansion joints in structures, should be made so that undue friction cannot develop between mating surfaces. Shouldered bolts, with a washer bearing on the shoulder after tightening as shown in Figure 3.4, may be used in slotted hole connections. In other bolted assemblies lock nuts will prevent the assembly becoming loose when in operation.

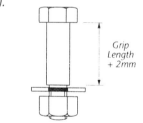

Figure 3.4 A shouldered bolt used at expansion joints

Machining operations are the focus of **3.4.10**. but they are rarely needed in normal building construction. For example *Joints in simple construction, Volume 2 (Ref. 2)* suggests that most base plates have a sufficiently flat bearing surface without machining or cold pressing. When the design does call for machining of a surface, suitable notes have to be shown on the drawing to direct the workshop's attention to the requirement.

The purpose of **3.4.11** is to identify where drilled holes, rather than punched holes are required. Its aim is to ensure that punching does not take place where, for example, distortion of the surface must be avoided. Work-hardening of the metal surrounding the hole could also initiate a brittle fracture in fatigue situations, or affect plasticity in areas where plastic hinges are required. Design codes limit the thickness of metal to be punched to the maximum values stated in 4.6.3.

The detailer must show on the Fabrication Drawings where holes are to be drilled and not punched.

In calculating the capacity of HSFG connections, **3.4.12** allows a slip factor of 0.45 for any clean steel friction surface **(faying surface)**, which is free from loose millscale, whether or not it has been blast cleaned. Such surfaces should be masked whilst protective treatments are applied to the surrounding areas.

For other surfaces the slip factor has to be determined by testing in the manner described in BS 4604 Part 1.

The NSSS does not preclude surfaces being roughened with a needle gun or by some other means to achieve a higher slip factor. Such treatments are sometimes needed on galvanized surfaces.

3.5   ERECTION DRAWINGS

The philosophy of the NSSS is that all drawing information needed by the erector should be found on purpose-made Erection Drawings. Although the erector will have a complete set of fabrication drawings on site, he will not normally refer to them unless he has occasion to check the fabrication details of a particular component.

The Erection Method Statement is discussed more fully in the notes to 8.1, but Erection Drawings, showing the erection method in outline, are expected to accompany the statement.

The drawings should also show the detail and arrangement of any temporary works or attachments to be used during the erection for lifting, restraint, propping or any other purpose.

3.6 DRAWING ACCEPTANCE

Acceptance or approval of drawings in **3.6.1** by the Engineer is a feature of most contracts and is a mandatory requirement of the NSSS.

The Engineer, having overall responsibility for the design of the structure, must ensure that the design information has been correctly interpreted on the fabrication and erection drawings. The clause is intended to apply equally to 'design and build' contracts when the designer is an employee of the Steelwork Contractor and is responsible for the steelwork design. His duties in relation to drawing acceptance are the same as those of a Engineer not employed by the Steelwork Contractor. Time must be allowed for drawing acceptance in the programme *(see notes to 1.8 (ii))*.

3.6.2 clarifies the **meaning of acceptance** for the purposes of the Specification. The Engineer, when accepting the drawings, must satisfy himself that the principle used in the design of a connection is correct, and it is the kind of connection he envisaged when designing the structure. If necessary, he should ask to see the design sheets for major connections.

The Engineer does not have to check the arithmetic used in designing the connection; that responsibility remains with the Steelwork Contractor, as does the responsibility for the accuracy of dimensions. Any mistakes resulting from these errors are the responsibility of the Steelwork Contractor.

Uniformity in phrases used by Engineers when accepting drawings is not a requirement but **3.6.3** tabulates the more **common terms** used and defines the meaning.

3.7 AS ERECTED DRAWINGS

The intention is that a full set of drawings will be retained by the Employer to be used for reference throughout the life of the structure. The Employer should preserve them and pass them on to subsequent owners; this process is an integral part of the C(D & M) Regulations.

Commentary on Section 4
WORKMANSHIP - GENERAL

Modern fabrication shops need to process steelwork with speed and efficiency. The old traditional skills of templating and individual marking of hole positions, by scribing lines on steel, have largely given way to the use of automated and semi-automated (numerically controlled) plant for cutting and drilling operations. Section 4 specifies a standard of workmanship for the principal operations in the modern fabrication process.

Material preparation tasks are considered to include cutting, machining, drilling, bending, punching and cropping. This is followed by assembly which can involve both welding and bolting; these procedures are covered in Sections 5 and 6. Inspection is a continuous process throughout fabrication, but a final inspection of completed components is an essential requirement.

4.1 IDENTIFICATION

The objective of **4.1.1** is that the steel used is **traceable** to a mill test certificate showing the manufacturing process, and mechanical and chemical properties, before the fabrication process commences.

There is no requirement for the Steelwork Contractor to have a system such that material can be traced to a particular certificate once the fabrication process has started; it is not a normal requirement for statically loaded structures. If continuous traceability is required, it would have to be written into the Project Specification, but it should be understood that it is an arduous task and will add to cost.

All **material grades** must be identifiable, but **4.1.2** takes account of the fact that the minimum sub-grade available for sections, plates and structural hollow sections can be taken when no special identification is shown. Higher sub-grades, per BS EN 10025, and grade S355 steel are usually identified by a stripe colour code system, painted at the end of the member on the cut edge.

4.1.3 requires the **erection mark** to be identified on the component during fabrication. Proprietary paint-sticks are often used for this purpose as chalk is not sufficiently durable.

Hard stamp marking of components is often one of the procedures carried out by automatic cutting and drilling lines. It is a good method of member identification throughout the fabrication and treatment process, but the indentations can form a stress raiser. The Engineer may need to indicate on his drawings any highly stressed tension zones in which hard stamping is to be avoided.

4.2 HANDLING

Steelwork is more likely to be damaged in storage and handling than in use. It cannot, by its nature, be cocooned in protective materials. The provisions of the NSSS are meant to ensure that damage from corrosion or handling is kept to a minimum by the use of appropriate slings and skids, and that components are stacked in such a way that sequential deliveries to site are facilitated.

4.3 CUTTING AND SHAPING

4.3.1 intends that **cutting operations** for all steel are performed accurately by mechanical means, but the Steelwork Contractor is free to use any one of the processes listed. However, it is recognised that machine guided flame cutting is not always possible and hand held equipment can be used in these circumstances.

4.3.2 requires a **flame cut edge** to be cleared of metal dross, and be free of indentations and irregularities. This may reasonably be interpreted as meaning indentations greater than one millimetre. Any greater imperfections should be removed by grinding or chipping. A local deviation of line which is better than 1:100 is considered reasonable. Although not stated, it would be expected that a flame cut edge should be within the tolerance of $t/_{10}$ given in 7.3.4

Columns are given special attention in **4.3.3**. Where direct end to end-bearing at a splice is not required, a greater cutting tolerance is permitted in accordance with 7.2.2. Where the ends are required to tightly butt to transfer axial forces, it must be remembered that joining the surface will also affect the plumb and alignment of the member. In these circumstances, the column ends must be cut to the more accurate tolerance in 7.2.3 taking care to prepare the end normal to a true centre line of the member *(Fig. 4.1)*.

Sawing of butting column lengths should be carried out, as far as possible, so as to reduce the effect of any natural curve in each member. If two members are sawn such that the ends complement each other, then the intermediate fabrication tolerance is unlikely to be of significance, provided that other tolerances are met once the column is erected. Where ends of columns sit on a base plate which is to be finally grouted, then clearly the end may be cut to the more generous tolerance in 7.2.2 *(Fig.4.1)*.

When very large columns with plan dimensions of more than one metre have to be in direct bearing, the 7.2.3 tolerance is considered to be too wide to ensure that erection tolerances can be achieved. Suitable conditions may be

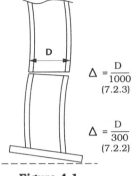

$$\Delta = \frac{D}{1000}$$
(7.2.3)

$$\Delta = \frac{D}{300}$$
(7.2.2)

Figure 4.1

obtained by matching butting components in the fabrication shop or by careful machining. 4.3.3 requires that these components are fabricated in such a manner that, after erection, tolerances in plumb and bearing are not exceeded.

In any bearing situation, notice must also be taken of the flatness requirement in 7.3.5.

4.4   MACHINING

The intention is that due allowance for machining is made when ordering plate and slabs so that the required dimension can be achieved after machining. In the case of base slabs, it is only necessary to machine a strip across the slab to suit the width or depth of the column section. (See *also notes to 3.4.10)*

4.5   DRESSING

Dressing of edges is not usually needed. Modern sawing, drilling and punching can all leave a satisfactory edge. Flame cutting can also be remarkably true. However, **4.5.1** requires arrised edges to be removed and holes cleared of **burrs** and protruding edges.

Sharp edges cut to an acute angle should be dressed so that they are not a danger in subsequent handling. A 2mm "rounding" would be acceptable. **4.5.2** states that a 90° rolled, cut, or sheared edge is acceptable without further treatment. However, it must not be a ragged or serrated edge.

4.6   HOLING

The NSSS expects holes to be positioned to tolerance, and true in alignment. Elements can then be assembled without distortion. Automatic numerically controlled plant is normally designed to ensure that proper account has been taken of deviations in flatness or straightness of components when holes are drilled.

4.6.3 stipulates the rules for good practice in **punching** and limits its use to material of 20mm thick or less depending on the steel grade. A further stipulation is that the thickness of material punched is not greater than the hole diameter. Positions in components where holes must be drilled and not punched, are specified in 3.4.11. Such positions must be drawn to the attention of the Steelwork Contractor in the Project Specification, as required by 1.3.1 (xv).

4.6.4 recognises that punching is permissible in any location where reaming is to take place. This is because any brittle hardening of the material around the edges of a punched hole does not extend beyond 2mm of the edge of the hole and the reaming operation removes this.

Slotted holes are dealt with in **4.6.5**. They may be produced using punching, drilling or flame cutting techniques but it should be noted that the punching criteria given in 4.6.3 still apply. Flame cutting of a hole should not have an irregularity greater than 1mm.

4.7   ASSEMBLY

Section 7 specifies permitted deviations for each element. An assembly may incorporate the permitted deviation in either direction *(see fig. 4.2)*.

If each element in an assembly is made to a tolerance of the same 'sign' (+ve or -ve), then the final assembly may be out of tolerance. To guard against this, the Steelwork Contractor must detail his components to avoid an accumulation of deviations which would make the final assembly unacceptable. Additionally, welded assemblies need to have balanced weld sequences to avoid warping and twisting.

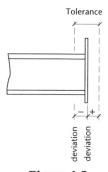

Figure 4.2

Steel is a tough, ductile, tolerant material, which forgives fairly rough treatment, nevertheless, the NSSS states that components must not be bent or twisted to achieve an assembly.

Refer also to 6.2 and 6.4 for fit-up of bolted assemblies with ordinary and HSFG bolts, and 5.4.1 for welding fit-up.

4.8   CURVING AND STRAIGHTENING

This clause is intended to take account of all fabrication operations where techniques are used to achieve the straightness or camber of a component within the permitted deviation given in 7.2.4, 7.2.6, 7.4.8, 7.4.9 and 7.5.8, and the curving of a component to a specified radius *(Fig. 4.3)*.

Straightening is within the capabilities of most steelwork contractors and some are able to undertake cambering. Bending to a specified shape is usually left to a specialist contractor.

The specialist can accommodate bending on the strong axis of all sections including the heaviest rolled. Castellated sections can also be curved.

Curving other shaped sections need not give the specialist any problems, but particular techniques have to be adopted to maintain the geometrical shape of both hollow   and open sections. For example RHS section can become slightly trapezoidal, and open sections may change slightly *as shown in Figure 4.4*. Generally the slight change of shape makes no difference to the strength of the component and acceptance or rejection should be on the basis of what is aesthetically acceptable.

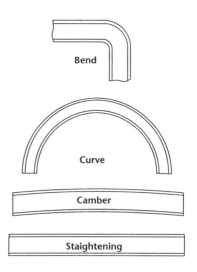

Bend

Curve

Camber

Staightening

Figure 4.3

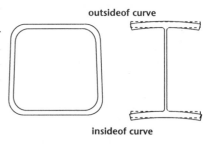

outsideof curve

insideof curve

Figure 4.4
Sections after curving

Most curving and straightening is generally performed by specialists on cold metal without heat being applied. The operations are in fact similar to the later stages of the final manufacture of rolled shapes. With cold working the mechanical properties of the metal will be changed locally. The yield and ultimate stress will increase whereas the charpy value and elongation properties will reduce. However, the component will be able to operate at normal stress levels when used in static building structures. Special attention may have to be given when fatigue loading is involved or where the structure may be subjected to very low temperatures.

Cambering and straightening by non-specialist steelwork contractors may be done by local application of heat along the length of one side of the member. Heating to a maximum of 650°C is not detrimental.

Induction bending is a relatively expensive process which may be the only way to bend very large sections and is particularly applicable to large tubes and rectangular hollow sections when the maintenance of a true cross section is critical. Temperatures in the material can be in the order of 800°C to 1000°C and must be very carefully controlled by trained operatives. A very high standard of curving can be achieved with no detrimental effects to the properties of the finished component.

In all these processes the component being straightened or curved is subjected to very large forces. Any welds made before straightening or curving must obviously be re-inspected and examined in accordance with 5.5.2, 5.5.3 and 5.5.4 after straightening or curving operations have been completed.

4.9   INSPECTION

Where automated (numerically controlled) methods of production are in use it should suffice to fully check the first components produced and the first batch of fittings. After that an occasional check of the components should be sufficient to prove that accuracy is being maintained.

The documents and procedures data sheets used in automated production should also be checked to the same standards as fabrication drawings.

If individual or small batches of components are being fabricated by traditional plating methods, with individual tape measurements being made to achieve the length, and holes drilled to scribed lines, then they should all be fully checked after fabrication.

4.10 STORAGE

It cannot be overemphasised that handling is not only expensive in itself but also carries the additional risk of damage, particularly to finishes. Nevertheless, production schedules are such that members, almost invariably, have to be stored at least once, either in the manufacturer's yard or on site. The purpose of **4.10.1** is to ensure that every effort is made to keep steelwork that is **stacked** clean, free from water collection and ensure that it is adequately supported.

4.10.2 requires that components should be marked in such a way that **visual identification** is possible when stacked. Sometimes proprietary adhesive labels are used for this purpose.

Commentary on Section 5
WORKMANSHIP - WELDING

A recognised specification providing weld acceptance criteria for steelwork in building structures was not available until the advent of the NSSS. Contract documents often contained specific requirements more appropriate for bridges and structures subject to dynamic loading. These requirements are usually more severe than are necessary for building structures.

BCSA commissioned The Welding Institute to set a minimum mandatory scope of weld testing, with acceptance criteria appropriate for static structures based on fitness for purpose. Cyclic stressing at fairly low amplitudes and frequencies does occur in building structures so the permissible size of discontinuities in welds were specified to ensure that they would not grow to failure during the life of the structure. The result of the work was Table 1 on Scope and Table 2 on Acceptance Standards. Welding procedure requirements and testing, and welder qualifications, relate to the requirements of Tables 1 and 2.

The aim is to ensure welds of adequate quality, by prior qualification of procedures and welders. In particular, the intention is to avoid or minimise repair work which is expensive, disrupts schedules and may be technically less acceptable. A guide to welding and effective quality control can be found in *Introduction to Welding of Structural Steelwork (See Ref. 5)*.

Care is needed in the use of welding terms. Sometimes expensive 'full penetration' welds are specified when what is needed is 'full strength' welds. A definition of welding terms is provided in the NSSS, but the following should be noted:

Fillet welds: The greatest proportion of welds in building structures are fillet welds made by manual metal arc (MMA), metal inert gas (MIG), or by the metal active gas (MAG) process. The weld fuses the joining parts, with a penetration into the parent metal of both pieces being joined and into the root. Provided that such welds are symmetrically made with the correct consumables, then they develop *'full strength'* when the sum of the weld throats is equal to that of the element which the welds join.

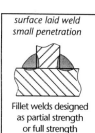

surface laid weld
small penetration

Fillet welds designed
as partial strength
or full strength

Full Penetration Welds: This weld has complete root penetration. It is used particularly in situations where the connected members have to accommodate a fluctuating load which could cause a fatigue failure in the weld. They are by definition always *'full strength'* but are considerably more expensive than other full strength welds.

Partial Penetration Welds: Butt welds which have an un-welded portion at the root are classed as partial penetration welds. The term also applies to a weld made from one side of the thickness where there is no backing and full penetration is not achieved. When partial penetration welds with a reinforcing fillet are symmetrically made with the correct consumables, and the sum of the weld throats is equal to that of the element which the welds join, they may be classed as *'full strength'* .

A butt weld to a circular or rectangular hollow section may be *'full strength'* when it is a continuous weld around the section and special techniques are used to ensure penetration through the wall thickness.

Full Strength Weld: The term does <u>not</u> describe the type of weld. It means that the joining welds must have a load carrying capacity equal to that of the element being joined.

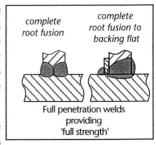

Full penetration welds providing 'full strength'

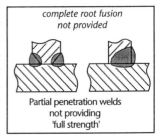

Partial penetration welds not providing 'full strength'

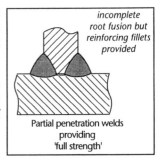

Partial penetration welds providing 'full strength'

5.1   GENERAL

This clause specifies arc welding processes to BS 5135: *Specification for arc welding of carbon and carbon manganese steels* (which is applicable to steels of maximum carbon equivalent of 0.54) and BS 4570: *Specification for fusion welding of steel castings.* BS 5135 provides a baseline for good workmanship and encapsulates a wealth of experience on techniques for producing welds of good quality by material selection and weld procedure specification. It also gives requirements for the use and storage of all welding consumables.

Welding consumables should remain in controlled conditions which avoid hydrogen pick up. Certain electrodes may need to be baked in suitable drying ovens prior to use and particular attention should be given to site conditions. Electrodes which are made and specially packed and sealed to avoid hydrogen pick up, may be advantageous on site and in similar environments, provided they are used within a given period after the package seal has been broken.

Consumables are chosen to ensure that the weld metal deposited has mechanical properties which are at least equal to those of the parent metal. These are commonly Grade E43 to BS 639 for welding S275 steel and Grade E51 to BS 639 for welding S355 steel. In fact BS 639 has been replaced by BS EN 499 but the cross reference in BS 5135 to BS 639 is applicable in principle to BS EN 499 (see also notes to 2.7).

Satisfactory welds will not be achieved unless the surfaces at the joints are clean and dry. Lightly rusted surfaces can be wire brushed, but deep rust must be removed by grinding or grit blasting.

Weld-through priming paints are acceptable provided that it has been shown by procedure tests, in accordance with BS 6084, that they do not have a detrimental effect on weld quality. In general, however, priming paint should be removed locally before welding.

5.2   WELDER QUALIFICATIONS

The NSSS requires that all welding is made by welders who have been properly tested and have certification to show that they have been tested for the type of welding they are undertaking. This is necessary since the quality of welded work performed by manual or semi automatic processes depends essentially on the skill of the welder.

Testing in **5.2.1** refers to BS EN 287: Part 1 *Approval testing of welders for fusion welding: Steels* . The testing method is one in which uniform rules and test conditions are specified using standard test pieces. However, the standard does not invalidate welder approvals made to other national standards or specifications, providing that the intent of the technical requirements is satisfied.

The same weld test can be used to test the welder, as well as approving the welding procedure to BS EN 288: Part 3, provided that all relevant requirements of both standards are satisfied.

BS 4872 Part 1 deals with the approval testing of welders for manual or semi-automatic welding where no approval of the procedure is required. The NSSS only permits this route for welders engaged in fillet welding.

It is required in **5.2.2** that testing of welders is witnessed and endorsed by an independent Inspection Authority. The welder's skill and job knowledge continues to be approved only if the welder is working with reasonable continuity on welded work which is covered within the extent of approval. BS EN 287: Part 1 provides a format for the certificate of approval.

5.3   WELD PROCEDURES

These clauses give the requirements for written specification of welding procedures and welding sequence. A welding procedure is defined as a specified course of action to be followed when welding, including a list of materials, and where necessary the equipment to be used.

It is the duty of the Steelwork Contractor to study the contract drawings/scope of the project to determine the range of welding procedure approvals required, and to review existing approvals and determine if any further ones are required. It is not necessary to have procedures re-approved for each contract when they can be shown to be part of a routine system of procedure tests operated by the Steelwork Contractor.

5.3.1 Requires **written procedures** to be available to BS 5135 and tested in accordance with BS EN 288: Part 3.

As stated in BS 5135, the following should to be recorded for such a test:

- the welding process;
- parent metal specification including dimensions;
- whether shop or site welding;
- dimensioned sketch of edge preparation;
- sketch of arrangement of run sequence and dimensions of the completed weld;
- method of cleaning and degreasing;
- make, brand and type of welding consumables;
- drying and baking of welding consumables before use;
- welding conditions;
- pre-heating and interpass temperature, including method and control;
- post-weld heat treatment.

Other items appropriate to the welding process will also be recorded, such as:

- travel speed (mechanised welding);
- current (a.c. or d.c.) and polarity for MMA welding;
- shielding gas and flow rate, nozzle diameter, arc voltage and wire feed speed for MIG or MAG welding;

Similar requirements apply also to other processes.

The principle in **5.3.2** is that **welding procedures and procedure tests**, made in accordance with BS EN 288: Part 3, are witnessed by a qualified Independent Authority. A Steelwork Contractor's own quality control staff ensure that the procedure is maintained, but approval comes from an outside body. There are a number of recognised authorities able to undertake procedure test approval.

Welding procedure sheets are intended to be the welder's instruction sheets and must be available at the work place whenever the work is to be carried out to written procedures. This is the objective of **5.3.3.** The sheets must also be available, on request, to any technical authority appointed by the Employer.

5.4   ASSEMBLY

The **Fit-up** specified in **5.4.1** is that relating to the joints themselves to allow the quality in Table 2 of the NSSS to be achieved. The examples shown in Table 2 illustrate the maximum permissible root gap with butting elements. The table also shows permissible tolerances in alignment of butt joints and cruciform joints.

Fit-up should also take account of the distortion anticipated during welding and the effect this will have on the final shape of the joint. Any presetting necessary to ensure correct geometry after welding can then be arranged.

The fit-up requirements for butt welds should be detailed in the weld procedures. Where correct fit-up cannot be achieved a build-up of weld metal by a "buttering" technique may be used to reduce a gap, if performed to an approved procedure.

Jigs are often used to ensure correct location in welded fabrications. Production rates may be improved by removing the workpiece from the jig after tacking. Providing that it can be demonstrated that adequate control of distortion can be achieved during subsequent welding, **5.4.2** allows removal from the jigs before welding is completed.

Tack Welding is a method used to hold assembled pieces until the main welds are performed. **5.4.3** gives mandatory rules for the sizing and quality of tack welds which are to be incorporated in main welds but such tacks must be made by welders who have qualified as detailed in 5.2. Any tacks which do not conform must be removed before main welding of that portion of the joint is carried out.

Tack welds have an increased propensity for cracking, because of the short weld runs and, hence, relatively low heat input. They may be highly stressed during removal from jigs and/or manipulation during welding. Welding conditions must, as a minimum, be in line with those for the root runs of the main welds.

It should be noted that when components being joined have a thickness of 12mm or greater, the tack weld must be at least 50mm long. When thinner members are being joined the tack length must be at least 4 times the thickness of the thinner member.

Distortion can be reduced by the adoption of the minimum amount of welding, an appropriate welding procedure and, hence, control of heat input. Weld deposits cause contraction to take place local to the weld. The amount of contraction is proportional to the amount of heat input from the welding process. **5.4.4** requires that weld procedures specify a sequence of weld runs which will balance the contraction forces to ensure the straightness of the components within the permitted tolerances in Section 7. In some cases, pre-bending or cambering may be necessary to counter weld distortion

To take care of longitudinal contraction in long welded components such as plate girders, the plates are ordered with sufficient allowance on the length to ensure that, after welding, the finished length can be achieved.

Welds, specified or made larger than necessary, do not provide stronger joints; they simply cause more distortion due to their greater heat input.

5.4.5 requires that the same attention is paid to welding **fabrication or erection attachments** as is made to permanent welds. Such attachments can be subjected to high stresses during lifting operations and can be dangerous if the welding is made without proper control.

3.4.4 states that these fittings can be left in place on the permanent structure, unless the Project Specification requires the Steelwork Contractor to remove them. If at all possible, they should be left in place.

Removal by hammering is forbidden since it can cause tearing of the surface of the member, creating stress raisers and leaving scars. If they are to be removed, the welds have to be carefully cut and grinding/buffing used as necessary. The remaining surface should then have the same level of visual and magnetic particle inspection (MPI), or dye penetrant inspection (DPI), as would be made for the weld metal.

Defects, such as cratering, may occur at weld start/stop positions. **Extension pieces** or run-on/run-off plates can prevent this situation. They are particularly useful when making in-line butt welds but are sometimes needed in other situations. **5.4.6** requires that the weld ends are dressed after careful removal of the extension pieces.

Where welded extension pieces are to be used as **production test plates** for mechanical tests, **5.4.7** states that the grade of material must be of the same specification as the parent metal. Rolled metal does have a 'grain' which follows the direction of rolling. The test piece must be placed in the same direction of rolling as that in which the parent metal was produced so that it can properly represent the quality of the weld in the joint.

5.5   NON-DESTRUCTIVE TESTING OF WELDS

The scope of inspection found in Table 1 and the weld acceptance criteria in Table 2 were developed for inclusion in a revision to BS 5950: Part 2, but this has been delayed owing to work on the European Standard EN 1090-1-1. It is currently expected that the UK National Application Document for the European Standard will refer to the tables in the NSSS. It is the intention that these will also be adopted for BS 5950 when it is next revised.

The quality system specified in Section 11 requires that records be kept of tests and inspections; **5.5.1** emphasises that **records** are to be made of weld testing. The clauses on non-destructive testing have been drafted as stand alone requirements which, together with the approved shop drawings, enable the Steelwork Contractor to inspect and correct the work in-house without referral to the Engineer for decisions. Only serious defects which are difficult to repair need to be referred to the Engineer for approval. However, all records have to be available for inspection by the Engineer.

The type of discontinuities which can occur in welds and their most likely causes should be understood. The inspection and measurement of defects can then be made with better interpretation and assessment of how important

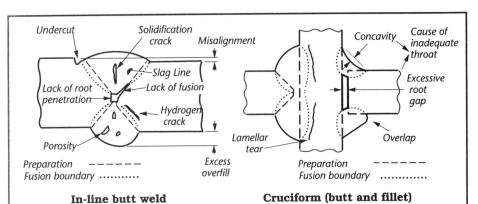

| In-line butt weld | | Cruciform (butt and fillet) |

Discontinuity Type	Most Common Location	Most Likely Cause
Hydrogen cracks	Heat affected zone (HAZ) in thick joints (can be in weld metal)	Damp or contaminated consumables or joint faces, too low heat input or preheat, high restraint. High carbon or Manganese in steel.
Solidification cracks	Weld metal (on centreline)	Deep narrow welds, high sulphur and phosphorus in weldment, high current. High restraint, mainly in submerged arc.
Lamellar tearing	HAZ in Tee and Cruciform joints	Non-metallic inclusions in steel, large welds, high restraint.
Lack of fusion	Side walls and root	Incorrect welding parameters, too low arc energy, travel speed too fast, incorrect electrode position.
Lack of penetration	Root	Incorrect fit-up, incorrect welding parameters (as per lack of fusion), inadequate back gouging.
Slag inclusions	Weld metal	Incorrect electrode manipulation, poor bead shape and inadequate slag removal, damaged electrode coatings.
Porosity	Weld metal	Gas originating from dampness, grease, rust, air entrainment in gas shield, insufficient deoxidant in consumable or steel.
Excess weld metal (overfill)	Weld bead cap	Excessive arc energy, insufficient preparation, incorrect electrode size, faulty electrode manipulation.
Undercut	Weld toe	Excessive current, too large a weld bead.
Overlap	Weld toe	Arc energy too low, travel speed too low, incorrect position, incorrect electrode manipulation.
Misalignment		Incorrect jigging, inadequate tacking.
Excess root gap		Incorrect jigging, inadequate tacking.

Figure 5.1 Common fabrication discontinuites and their causes

a defect is. Common fabrication discontinuities and their causes are shown in Figure 5.1 which is reproduced from M.H. Ogle's paper *Weld Quality Levels (See Ref.6)*.

Visual inspection of welds, at a level of 100%, is required by **5.5.2** before other flaw detection methods are used. This inspection is to ensure that the right type of weld is made in the right location and no weld is missing. The dimensions of throat and leg lengths of fillet welds, overfill and undercut, concavity and misalignment can be measured with the aid of a hand-held gauge.

Table 2 makes clear the remedial measures to employ if any discontinuities are present.

When welds will become inaccessible by later workmanship, it is necessary to perform such inspections and take corrective measures before they become inaccessible. BS 5289 provides useful guidance on visual inspection which is appropriate for the work being inspected and the type of weld.

The NSSS gives preference in **5.5.3 to Surface Flaw Detection** by magnetic particle inspection (MPI). This method works on the principle that a magnetic field in a piece of steel will leak out to the surface when a crack is present. The particles of iron powder are attracted to the crack location, producing a visual indication of a crack which might otherwise be invisible to the naked eye.

For carbon steels, magnetic particle testing for surface cracks is preferable to dye penetration testing methods. However, the NSSS allows the use of dye penetrants when MPI is not available.

BS 6072 *Method for magnetic particle flaw detection* gives advice on the materials to be used, safety precautions, testing procedures, surface preparation, magnetization, application of detecting material, viewing, recording, reporting and demagnetization.

The minimum requirement for the scope of testing is detailed in Table 1 of the NSSS, which advises the thickness of material for each weld type at which MPI becomes mandatory. For welded connections, the first five of any particular type of joint must be fully tested and one in five thereafter. In site conditions testing must be 100% of all joints identified as requiring testing by Part B of Table 1.

A reduced amount of testing is permitted for welding in zones of members which do not contain a connection. This is associated with components made of elements welded together, such as plate girders and web stiffeners. Welding to the haunch of a portal frame rafter may be considered as part of the component assembly and not of the connection. It is, therefore, only subjected to the reduced amount of testing as a "member zone" component. Welded transverse butts must always be treated as connections and given the higher frequency of testing.

The higher frequency of testing at the initial stages is set so that confidence can be gained in the procedures adopted. When welded joints have, for example, sudden changes in member cross section or high stress concentrations at openings in members, the Engineer should consider having a greater frequency of testing *(see also the notes to clause 1.4(vi)).*

A period of 16 hours between welding and testing for S275 (Design Grade 43) and S355 (Design Grade 50) steels is the <u>minimum time</u> which must elapse to ensure that if hydrogen cracking is going to occur, it will have done so. A 'next day ' approach to testing is therefore a necessity. If cracks appear due to an error in procedure, e.g. moisture contamination of electrodes or failure to provide proper preheat, it is likely to affect many of the welds in the batch. The prime purpose of the inspection is to spot the basic problem. Note that Table 2 requires a 100% MPI when a crack is found.

When the weld procedure requires the weld to be tested at an intermediate stage of the fabrication process, the work piece will have cooled and will have to be reheated to the temperature recommended in BS 5135 before welding continues.

When a welded component is highly restrained, or when welding thick materials, say 50 mm and above, inspection should take place after a longer period of 40 hours as for S460 (Design Grade 55) steels.

Competence in surface flaw detection can be gained by a welder who has been suitably trained, and has been tested by a nationally recognised authority. It is acceptable for a welder who is so qualified to examine his own work.

In **5.5.4 Ultrasonic Examination** is specified for subsurface examination of welds and no mention is made of Radiography. Ultrasonic examination is favoured for all except thin material (up to 6-8mm thick) and small fillet welds. Radiography is slow and disruptive to other operations in the vicinity and carries severe risks to health and safety.

Ultrasonic examination is to be made in accordance with BS 3923. Part 1 covers manual methods of weld examinations whilst Part 2 concerns automatic examination of fusion welded butt joints. Various levels of examination of in-line butt welds are covered; the NSSS requires level 2B. Ultrasonic examination is subject to the same elapsed time requirements before testing as magnetic particle inspection.

All ultrasonic examination requires considerable skill and training. It must be carried out by operators who hold a current certificate of competence in making an examination and interpreting the results.

TABLE 1 WELDS - SCOPE OF INSPECTION

This table shows the minimum mandatory scope of inspection required by **5.5.5.** Part A stipulates that a 100% visual inspection is required, and Part B advises the material thickness, or fillet weld size, when MPI and Ultrasonic examination become mandatory. The commonly used steels in grades S275 and S355 have the same requirements, whilst the rarely used S460 steel is more stringent. Part C provides the requirement for frequency of testing.

It should be noted that where MPI and ultrasonic examination are specified they are **not** alternatives; both must be carried out since MPI can only discover surface discontinuities, and ultrasonic examination is required to check for sub-surface discontinuites.

In Part B of the Table the criteria for mandatory testing by MPI and ultrasonic inspection methods is based upon the thickness of the thickest element in the connection, and on the importance of the welded joint. It should be appreciated that the failure of some welds could be catastrophic, whereas others are less critical. It is matters of this nature which have influenced the decisions upon the minimum thickness of steel at which testing becomes mandatory as well as what is practicable.

Ultrasonic examination (U/S) of single sided welds can usually only take place from one side, whereas both sides of a double sided weld can be examined. Section B of the Table specifies ultrasonic examination being required for single sided welds when the plate thickness is 12mm or greater, and for double sided welds 30mm or greater.

In-line butt welds, which connect two members which carry a tensile force, must have a non-destructive test with all but the thinnest steel.

Tee, cruciform and corner butt-welds are often subjected to shear loads only. These are considered to be less critical. MPI and ultrasonic examination therefore commence with thicker steels only.

It is incumbent upon the Engineer to recognise where a particular connection may require non-destructive testing of welds to steels which are thinner than those specified in Table B; for example, a cruciform weld situation where there is an in-line tensile force to be carried through the parting plate. In this case testing may be necessary for steels as thin as 10mm, with ultrasonics being applied to the plate as well as the weld.

Since joints for testing are related to material thicknesses they can be recognised easily once the fabrication drawings are made. One way of scheduling the weld testing of similar components (which may be shown on a batch of drawings) could be as illustrated in Figure 5.2 for portal frame haunch connections and welded purlin brackets. A total of 40 rafters with 20mm thick flanges and a connection with a 20 thick end plate is assumed, to show the quantity of MPI which would be required.

Table 2   WELD QUALITY ACCEPTANCE CRITERIA & CORRECTIVE ACTIONS

This table shows the acceptance criteria and corrective actions required by **5.5.6.** Since the table is based upon 'fitness for purpose' criteria, a lesser standard is not permitted. The welding standards specified are achievable by any competent Steelwork Contractor.

The clause indicates once again that the acceptance criteria are for structures subject to static loading only. In this context wind loading is considered as a static load unless the wind can set up a frequency of oscillation similar to the natural frequency of the component. In general this is only likely in certain tubular structures, towers and chimneys. (See BS 4076 *Specification for Steel Chimneys* and BS 8100 *Lattice Towers and Masts.)*

If the designer specifies that fatigue loading is to be applied to parts of a static structure, such as an overhead crane gantry, or a building housing reciprocating machinery, it is recommended that weld acceptance criteria should be in accordance with the bridge specification BS 5400 Part 10 *Code of practice for fatigue* or BS 7063 *Fatigue in steel structures.*

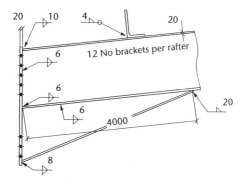

Weld Inspection Sheet			

For *Portal Frames* __ **Drawing Number** *C 8250/34, 35, 36 and 42*

Weld		Visual Inspection	M.P.I.
Location	Type		
40/ Haunch Tee	6 FW	320 metres	16 metres
	20 FW	12 metres	3.6 metres
40/ End Plates	10 FW (top)	24 metres	7.2 metres
	6 FW (bottom.)	24 metres	7.2 metres
	8 FW (haunch.)	24 metres	7.2 metres
	6 FW (web)	88 metres	26.4 metres
Purlin Brackets (40 x 12 sets)	4 FW	240 metres	12.0 metres

40 x 4 metres x 2
320 ÷ 10 = 32 x 0.5

40 x 0.3 metres
5 + 35 ÷ 5 = 12 x 0.3

40 x (2 x 0.3) metres
5 + 35 ÷ 5 = 12 x 0.6

40 x (2 x 0.3) metres
5 + 35 ÷ 5 = 12 x 0.6

40 x (2 x 0.3) metres
5 + 35 ÷ 5 = 12 x 0.6

40 x (2 x 1.1) metres
5 + 35÷ 5 = 12 x 2.2

40 x 12brkts x 0.5
40 x 12 ÷ 20 = 24 x 0.5

Figure 5.2
Typical Weld Inspection Sheet

The acceptance criteria are not difficult for a competent welder to achieve, but **Table 2** needs careful study in order for it to be fully understood. The following notes are made to assist in the interpretation of the table.

Weld Geometry

The requirement is that the weld location, type and length should be as shown on the fabrication drawings. If repairs have to be made because the weld has been wrongly located, an approved procedure has to be used and the inspection requirement for the repair is increased to 100%.

If a wrong type of weld has been made. e.g. if a fillet weld had been used because the preparation groove for a partial penetration weld was forgotten, then the repair procedure would have to be referred to the Engineer.

Profile

Throat thickness 't_b' or 't_f' must *average* the throat indicated by the drawings over each 50mm length of weld. It shall be not greater than 't_b' + 5mm, or 't_f' + 5mm at any point.

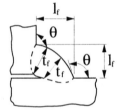

Leg Length (fillet) Usually l_f is required to be equal for both sides of the fillet, but in special cases the designer may specify different leg lengths on the drawing.

Toe angle, θ, of any weld must not be less than 90°.

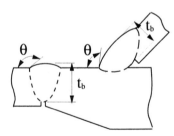

If repair or grinding is necessary to keep within tolerance, an approved procedure is to be used, and grinding repairs inspected by 100% MPI.

Cap/Root Bead The discrepancy 'C_b' in the height or concavity of a weld bead must be between the limits −1mm and +4mm.

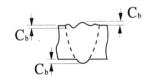

Repair to an approved procedure or grind as necessary.

Misalignment The discrepancy 'm' in the alignment of an in-line butt joint shall not be more than a quarter of the thickness 't' or 3mm, where 't' is thickness of the thinner member.

The discrepancy 'm' in the alignment of a cruciform joint shall not be more than half of the thickness 't' or 3mm, where 't' is thickness of the thinner member being joined.

If the discrepancy is more than these amounts, it must be referred to the Engineer for a decision on acceptance or repair.

Surface Discontinuities

Undercut   For a transverse weld, the sum of the undercut '$U_1 + U_2$' must not average more than 1mm or '0.1t' over each 100mm of weld, where 't' is thickness of the thinner member being joined. In practical terms this means that no measurable undercut is allowed when welding thin material. However, in the case of longitudinal welds, as defined by Table 1, these limits are doubled allowing small undercuts in thin material.

Repair to an approved procedure.

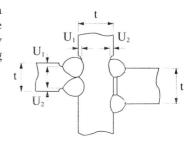

Lack of Root Penetration for <u>Single sided butt welds</u>

A single sided butt weld is commonly used in welding a hollow section onto another component, and in these cases, the root of the weld cannot be inspected. For in-line single sided butt welds to hollow sections, which are at least 10mm thick, an ultrasonic test must be made to establish the weld thickness.

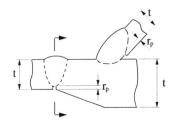

The illustrations show examples where the intention is that a weld is required with a throat equal to that of the member thickness. The value r_p is the permitted deviation from that requirement.

For a transverse weld, as defined in Table 1, the effective weld throat lack of root penetration 'r_p' must not exceed 3mm or average more than 0.05t over a 100mm length of weld, where 't' is thickness of the thinner member being joined. However, for a longitudinal weld, whilst the 3mm maximum is maintained, a relaxation is permitted to 0.1t.

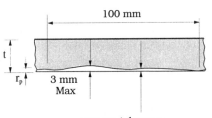

average tolerance
0.05 t for transverse welds
0.1 t for longitudinal welds

Repairs must be carried out to an approved procedure and where the root is inaccessible, the amount of ultrasonic inspection is doubled.

Surface Porosity.

The size of surface cavities 'd' shall not exceed 3 mm and in a transverse weld, the sum of 'd' shall not exceed 10 mm in a 100 mm length. However, in a longitudinal weld, the sum of 'd' shall not exceed 20 mm in a 100mm length.

Repair to an approved procedure.

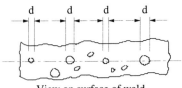

View on surface of weld

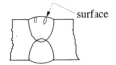

Typical Discontinuities

Lack of Fusion and Cracks

Lack of Fusion and Cracks are **NOT PERMITTED.** Any found must be repaired to an approved procedure.

MPI is increased to 100% after repair.

Sub-Surface Discontinuities

Root penetration

In butt welds which are not intended to provide full penetration, lack of fusion (or the gap 'h' remaining after root penetration) shall not exceed the amount shown on the drawings plus 3mm.

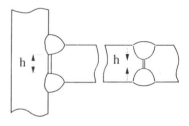

When elements are prepared with a nosing intended to achieve partial penetration, with or without a superimposed fillet in accordance with BS 5950, the preparation should be made as shown here.

Repair to an approved procedure and double the scope of ultrasonic examination.

Full details on the design of partial penetration welds may be found in the section dealing with end plate connections in *Joints in Steel Construction - Moment Connections (Ref. 7)*

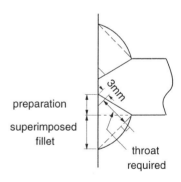

preparation

superimposed fillet

throat required

Slag Lines. If within 6 mm of the surface, or of another slag line, the thickness 'h' shall not be greater than 3mm, and slag lines must not exceed 10 mm in length. Longitudinal separation of slag lines must be equal or greater to 10mm.

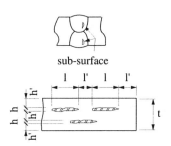

sub-surface

Typical section through sub-surface defects.

In addition to above, in a transverse weld, as defined in Table 1, the sum of 'l' shall not exceed 3t(mm) in a 200mm length. However, in a longitudinal weld, the sum of 'l' shall not exceed 6t(mm) in a 200mm length.

Repair to an approved procedure and double the scope of ultrasonic inspection.

Root gap. In fillet and partial penetration butt welds, used in 'tee', cruciform, corner and lap joints, the root gap $'r_g'$ shall not exceed 3mm, or 2mm average in a 100mm length.

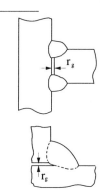

Repair to an approved procedure and double the scope of ultrasonic examination.

Sub-Surface Cracks As with surface conditions, cracks are **NOT PERMITTED.** Any found must be repaired to an approved procedure and ultrasonic examination increased to 100%.

Lamellar Tears Lamellar tears are **NOT NORMALLY PERMITTED;** refer to the Engineer for a decision and increase ultrasonic examination. An exception may be made for welds having to develop longitudinal shear only.

Lamellar tearing is mainly due to cracking, where non-metallic inclusions are present in the steel, occurring under shrinkage forces of welds. There are techniques for overcoming the problem; expert metallurgical advice should be sought.

5.6   SHEAR STUD WELDING

The use of shear studs, which is extensive in steel/concrete composite construction, is not covered by a British Standard although a European Standard is in the course of preparation. The NSSS therefore covers trial welding, tests and inspections and replacement of defective studs, for conventional use in building structures.

One technique is to weld the stud, on site, directly through gauge metal decking onto the steel beam below, but studs used with precast floor units are often shop fixed.

The equipment to be used and procedures must be as recommended by the equipment suppliers, particularly as the work may be carried out on site after erection of the steelwork,

Trial Welding in **5.6.2** is only stated as being required when specified by the Engineer, but it is possible that the Steelwork Contractor could be testing on each shift.

Where the studs are to be welded through a primed paint surface, such priming paints must be applied to the trial work pieces. However, it may well be necessary to remove the primer paint to achieve a good stud weld.

The **bend test** given in **5.6.3** should be carried out with the first shear connections made to prove that the weld procedure used and the welding current was correct. Further tests should be carried out at intervals to achieve a total of 5% of studs tested. A typical bend test programme is as shown in Figure 5.2. It should be noted that the 5% studs which are bend tested should not be straightened again.

3	*tested in first*	10
then 3	*tested in next*	10
then 2	*tested in next*	80
then 2	*tested in next*	100
10	*total in*	200

Figure 5.2
A typical bend test
programme for shear studs

It is required by **5.6.4** that **defectively welded studs** are removed carefully as if they were temporary attachments, hence the reference to 5.4.5.

Commentary on Section 6
WORKMANSHIP - BOLTING

The logic behind these clauses in the NSSS is that bolting should be kept as simple as possible. Ordinary bolted assemblies should therefore be considered as the preferred system for building structures. The use of 8.8 fully threaded bolts is recommended as discussed under 2.8.

The debate about making site connections by welding instead of bolting is likely to become more intense in the next few years. Riveting, which was so popular in the early days of the structural steelwork industry, finally gave way to bolting; it has yet to be seen just how far the change from bolting to welding on site will go.

Bolted connections are popular with Steelwork Contractors who have automatic sawing and drilling plants where holing operations can be carried out without having to individually mark the members. Connecting cleats are often made by cropping angles and punching holes.

As stated in the notes to other Sections, steelwork in buildings is mainly subjected to static loading, but vibration can occur in some situations. In these circumstances provision has to be made to prevent bolts from working loose. No particular guidance is given in the NSSS but the Engineer must give consideration to lock nut devices such as those in BS 4929: Part 1, or other means to prevent the bolt from working loose, such as the use of preloaded bolts which do not necessarily need to be high strength friction grip (HSFG). However, in practise, HSFG bolts are nearly always used when preload is required. Equally, when connections will be regularly subjected to a reversal of the direction of the load, it is the Engineer's responsibility to choose a suitable type of bolt assembly.

6.1   ORDINARY BOLTED ASSEMBLIES

The **Bolt/Nut Combinations** in **6.1.1** are those intended for use in shear or tension situations. 10.9 bolts with grade 12 nuts have been introduced into the 3rd Edition of the NSSS; they are commonly used in some European countries.

Care should be taken when using 10.9 bolts in tension owing to their limited elongation capacity. It is recommended that the bolt length is chosen which ensures that at least five threads will be available under the nut after tightening.

6.1.2 precludes the use of **differing bolt grades**, of the same diameter, on the same job as a safeguard against the weaker bolt being used in a situation where the stronger bolt is necessary. It is good practice use 8.8 bolts for main members and to limit the use of 4.6 bolts to secondary members such as sheeting rails and purlins where a different diameter can be adopted.

The **bolt length** requirements in **6.1.3** should be regarded as the minimum for good practice. It should <u>not</u> be quoted as a reason for having lots of different length bolts in use. The clause can be equally satisfied by using a standard length fully threaded bolt as discussed in the notes to 2.8.

It is permissible for the threaded portion of the bolt shank to be within the shear plane of a connection and load capacity tables for bolts are based on this assumption in BS 5950. A paper by G W Owens gives the background to fully threaded bolts in both tension and shear conditions *(Ref. 8)*.

6.1.4 does not require **washers** to be used with ordinary bolt assemblies. However, washers are required under inclined flanges of joist sections, where protection is needed from the rotated nut or head next to a surface with a finished protective treatment, and when bolts are used in oversize or slotted holes.

Nuts in **6.1.5** which are hot **galvanized** cannot be expected to have a zinc coating which follows exactly the thread profile. It is therefore necessary to check the nut for free running on the bolt and re-tap when it is not free running.

The intention of **6.1.7** on **bolt tightening** of ordinary bolts is that the bolts are at least 'spanner tight' whether they be assembled using impact tools or hand spanners to BS 2583. This may be defined approximately as

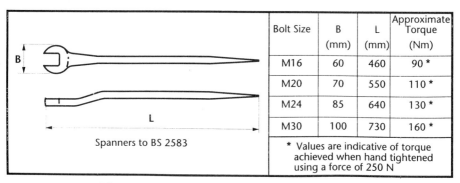

Bolt Size	B (mm)	L (mm)	Approximate Torque (Nm)
M16	60	460	90 *
M20	70	550	110 *
M24	85	640	130 *
M30	100	730	160 *

Spanners to BS 2583

* Values are indicative of torque achieved when hand tightened using a force of 250 N

Figure 6.1 Bolt tightening with podger spanners

being at least a 25kgms (250 Newtons) force being applied to the end of a podger spanner. *See Figure 6.1.*

Fitted bolts are sometimes used in non-slip situations as an alternative to HSFG bolts. **6.1.8** gives the requirements when precision bolts are used. The holes in the components to be joined can be individually drilled undersize and then reamed so, when assembled, the bolt has no more than 0.3mm clearance. Alternatively, where possible, they can be drilled full size through two or more thicknesses in one operation.

6.2 FIT-UP WHEN USING ORDINARY BOLTS

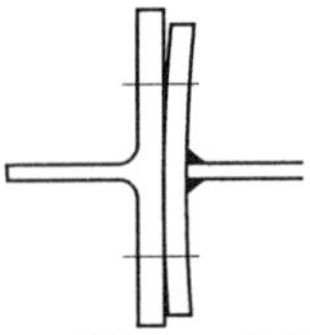

Although connected parts must **fit-up** tightly when drawn together, it is implicit in **6.2.1** that there may be small gaps since they are not machined surfaces. Packing is only necessary when the integrity of the joint is threatened. In a case such as that shown in Figure 6.2 there is clearly no loss of joint integrity even though the end plate has curled slightly in welding.

Figure 6.2
Gap not affecting joint integrity

Although it is common practice to drive a drift through matching holes to achieve alignment, excessive force causing distortion to steelwork is not acceptable in the NSSS. **6.2.2** requires that the holes should be **reamed** and a larger diameter bolt used to achieve a proper connection.

6.3   HIGH STRENGTH FRICTION GRIP BOLTED ASSEMBLIES

Preloaded bolts have advantages where vibration is present, where slip between joining parts must be avoided and when the applied load through the joint frequently changes from a positive to a negative value; otherwise, 8.8 bolts will prove more economical to use.

It is a fact that HSFG bolts had an early history of misuse on site. It is therefore important, when fixing, to have a higher level of supervision than for ordinary bolting.

For HSFG bolts, the requirements for bolt/nut/washer assemblies and for tightening are those demanded by the British Standards. A slip factor for a clean steel friction surface (faying surface) is provided in 3.4.12 of the NSSS; otherwise it should be determined by testing.

In **6.3.2** the use of the torque control method of **tightening**, or the part-turn method described in BS 4604 is permitted; proprietary load indicating washers, illustrated with applications of use in Fig 6.3 can also be employed. In the normal case the load indicator is placed under the head of the bolt and the nut is rotated on a hardened steel washer *(left-hand diagram)*. Should circumstances dictate, it is also possible to have the load indicator placed on the nut side with the nut rotated on a 'nut face washer' to prevent direct contact between the load indicator and rotating surfaces *(middle diagram)*. A further special case is when the head of the bolt is to be rotated to achieve tightening; in this case the load indicator and a 'nut face washer' is used on the nut side and the head is rotated with hardened washer under *(right-hand diagram)*.

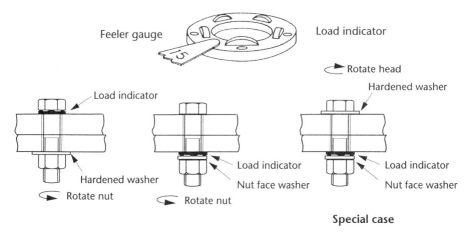

Figure 6.3 Use of load indicating washers with HSFG Bolts

The use of torque control methods to achieve a particular tension in a bolt is the most difficult to control since it depends upon the condition of the threads, the amount of lubricant present and the state of the nut/washer interface. Consequently **6.3.3** requires a **calibration check** to be made at least once per shift.

When a preloaded bolt assembly has been properly tightened the bolts have been stretched beyond the elastic limit of the material. If for any reason the bolt is then slackened off, **6.3.4** says the bolt assembly must be **discarded** as required by BS 4604. However, a bolt may be reused during erection when it has been only partially tightened, to not more than 80% of its proof load, since its properties will not have been impaired.

6.4 FIT-UP WHEN USING HIGH STRENGTH FRICTION GRIP BOLTS

6.4.1 requires that the joint be examined for **fit-up** when all bolts have been partially tightened, to ensure that the steel surfaces are properly in contact.

Lack of fit in HSFG connections is discussed at length in a CIRIA publication by A P Mann and L J Morris *(Ref. 9)*. They quote research suggesting that HSFG connections subject to shear should have the plate material adjacent to the bolts clamped together over an annular zone whose maximum diameter is about twice that of the bolts.

As in 6.2 above, a small gap on the edge of the meeting surfaces is not detrimental to the integrity of the joint. Only gaps caused by an actual geometrical problem need packing.

The requirements of **6.4.2** on **reaming** are similar to 6.2.2 for ordinary bolt connections. However, since the load is being taken in friction, it may not be necessary to use larger diameter bolts.

6.4.2 (ii) requires that calculations should be made to demonstrate adequate capacity in the connection for the reamed condition.

Commentary on Section 7
WORKMANSHIP - ACCURACY OF FABRICATION

When compiling the NSSS, consideration was given to omitting the section on accuracy of fabrication. The suggestion had been made that only accuracy of erected steelwork really mattered in terms of the contract, and how individual components were manufactured could be left to the Steelwork Contractor. However, steelwork for one project may be made in several different locations, with more and more components being produced on automatic or semi-automatic plant, needing clearly understood quality levels. It soon became obvious that a statement had to be made of the accuracy which could be achieved when using properly maintained equipment in a well ordered workshop. These are the standards, referred to as **permitted deviations** in **7.1**. They provide acceptance criteria for incorporation into quality control procedures at the time of fabrication.

A skeletal steel frame is there to support and stabilise the rest of the building structure. Accuracy of fabrication is therefore of concern to all dealing with the floors, walls, cladding, glazing and anything else fixed to or supported by the steel frame. Accuracy of steelwork is not critical in many cases, but in other instances it is of paramount importance and ignorance of Sections 7 and 9 can lead to untold costs and delays if provision is not made for the dimensional deviations which can occur.

Cognisance should be taken of likely deviations when preparing the fabrication drawings so that the most suitable detail can be chosen, or allowances can be made for adjustment. Packings may be necessary in order to align members and to prevent a long line of butting members 'growing', as small discrepancies build up. *(See 3.4.6.)*

7.2 PERMITTED DEVIATIONS IN ROLLED COMPONENTS AFTER FABRICATION

Cross Section After Fabrication

The starting point for accuracy in workmanship in steelwork construction is the dimensional tolerance of hot rolled sections which is determined at the rolling mills; this is the subject of **7.2.1**. The fabrication process is not intended to be used to improve the geometric accuracy of rolled components in cross-section. The principal tolerances permitted on the shape of cross-section of rolled components, taken from BS 4: Part 1 and BS 4848: Part 4, are provided earlier in *Fig 2.3* .

Squareness of Ends not Prepared for Bearing

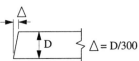

NSSS 7.2.2
Ends not in bearing

The squareness of ends of components is usually more significant in terms of accuracy when the member is in compression and bearing directly onto another member. In any case, some control is needed to ensure the correct length of the member. The deviation stipulated in **7.2.2**, when the end is **not in bearing** is considered satisfactory for members in simple construction; the more accurate requirement of **7.2.3** would normally be required for members where the load is taken in **direct bearing**.

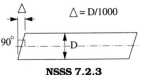

NSSS 7.2.3
Ends prepared for bearing

The reference to 4.3.3(i) is made to indicate that column shafts connected to slab bases, where grouting is employed under the slab base, do not require the greater accuracy in bearing (as long as the flatness of the upper surface of the slab is maintained).

Columns, and other components in bearing, which have cross-section dimensions greater than one metre may require the squareness of the end to be more exact in order to achieve accuracy in erection; hence the reference to 4.3.3(iii).

Care should be taken, when checking the deviation, to ensure that it is square to a true centre line which bisects each end of the member and not square to one end, otherwise any curvature in the member will be translated into a bearing surface which will not be true when the member is plumbed or aligned in the structure.

No reference is made in 7.2.2 or 7.2.3 to sawing or machining. The accuracy in **7.2.3** can usually be achieved by sawing when using a well maintained machine. However, machining may be necessary when an accuracy of a saw cut end would not be within the permitted deviation.

Straightness and Length

The **straightness tolerance (7.2.4)** is normally achieved by material delivered from the rolling mills and this requirement automatically meets the design criteria in BS 5950 Part 1. If a greater degree of accuracy is required a straightening operation will have to be carried out in the workshop. Correcting very small discrepancies can be expensive.

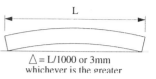

$\triangle = L/1000$ or 3mm whichever is the greater

NSSS 7.2.4 - Straightness

The deviation of one thousandth of the length was chosen because this parameter is within that normally assumed in buckling formulae when designing struts.

The **length deviation** of ± 2mm, found in **7.2.5,** is less generous than BS 5950 Part 2 which specifies ± 3mm. However, it should be noted, that in the NSSS the length is measured on the centre line of the member and does not take account of the squareness of ends.

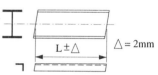

$L \pm \triangle$ $\triangle = 2$mm

NSSS 7.2.5 - Length

For **curved or cambered members 7.2.6** allows twice the minimum amount of deviation as 7.2.4 for straight members. Theoretically the curve should mirror the deflected shape which is usually parabolic, however, an arc of large radius or a curve over the centre portion will suffice. The deviation is measured at the centre of span only.

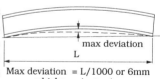

max deviation

L

Max deviation = L/1000 or 6mm whichever is greater

NSSS 7.2.6 - Cambered

Members curved for architectural purposes rather than to counteract deflections should have maximum deviation at centre of span with proportional deviations at other positions.

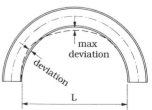

max deviation

deviation

L

Max deviation = L/1000 or 6mm whichever is greater

NSSS 7.2.6 - Curved

7.3   PERMITTED DEVIATION OF ELEMENTS OF FABRICATED MEMBERS

Fittings

The need for proper alignment when transferring load from one member to another was recognized when preparing **7.3.1**. A typical detail envisaged could be a bolted beam to column moment connection which requires a compression stiffener to reinforce the column as shown in Figure 7.1. The stiffener should not be more than 3mm out of nominal alignment with the beam compression flange.

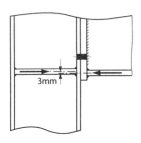
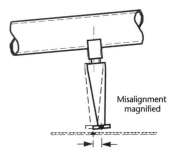

Figure 7.1 **Nominal alignment of a** **compression Stiffener**	**Figure 7.2** **Deviation magnified** **at connection**

Consider a screwed spigot boss as shown in Figure 7.2 welded to a roof member which is intended to receive a glazing support bracket. The glazing panel may be supported at four such locations. In this case the boss may be within 3mm of its true position but suffer an angular displacement which could magnify the positional discrepancy several times at the point of connection to the glazing. Where there is repetition, and several brackets locate on single components, the angular displacement can be retained in relative positions by a suitable jig, but the problem can still be present when a group of components have to provide a grid of similar connections. The best solution to suit all locations is to have a bracket connection which allows 3-dimensional adjustment; a recommendation which would apply to any interface between different materials and where responsibility for construction is shared *(see notes to 9.3).*

Holes

The permitted deviation of 2mm in the **position of holes**, as required by **7.3.2**, takes account of normal bolt clearances and avoids the possibility of material being distorted during bolting up. If holes are part of a group in the same connection then clearly the deviation must apply to the whole group in the same direction.

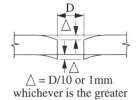

NSSS 7.3.2
Position of holes

Any significant distortion in the material around **punched holes** will prevent surfaces meeting properly. The permitted deviation in **7.3.3** can easily be attained by modern punching equipment. Where punching is employed in both elements being connected, it is recommended that punching is made in one direction to avoid doubling the gap at meeting surfaces.

$\triangle$ = D/10 or 1mm
whichever is the greater

NSSS 7.3.3
Punched holes

Surface Flatness

The degree of flatness for a full contact bearing surface envisaged for **7.3.5** was that the surface should appear flat to the naked eye when a straight edge is placed against it. If two mating surfaces are within this deviation, and the squareness criteria (7.2.3) is observed, it is normally possible to erect and align the joining members with a gap no more than that stipulated in 9.5.5.

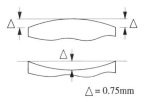

$\triangle$ = 0.75mm

NSSS 7.3.5
Surface flatness

7.4   PERMITTED DEVIATIONS IN PLATE GIRDERS

Most plate girders are manufactured using automatic equipment for plate assembly, with main flange-to-web welds made using continuous submerged arc welding techniques. Stiffeners are welded into position with other automatic, or semi automatic welding plant. Individual plates used for plate girders are mainly cut to width using automatically operated flame cutting equipment.

The permitted deviations given in the sub-clauses of **7.4** are those which are attainable with the equipment described, but some thin flanges may have to be preset to ensure their flatness after welding. A special requirement is provided in **7.4.11** for squareness of flanges to web at bearing support positions, so that the web properly resists vertical shear, and the permitted deviations of the web and web stiffeners are those which are assumed in normal out-of-plane web buckling design criteria.

The provisions for length, overall straightness and curving or cambering are essentially the same as for rolled members. It is envisaged that curved and cambered shaping is achieved by cutting the web plate to shape before welding, rather than bending the girder after welding.

7.5   PERMITTED DEVIATIONS OF BOX SECTIONS

Box girders, like plate girders, are likely to have plates prepared for welding by automatic plant. Permissible deviations are very similar to those for plate girders. Suitable edge and end preparations coupled with balanced welding operations should ensure that straightness requirements are met.

A box, being torsionally stiff, would be difficult to square up after welding. It is envisaged that squareness is retained by properly shaped diaphragms. The measurement of **squareness** is in **7.5.2** is made relative to the adjacent sides as, in practice, it is much easier to measure than a diagonal dimension.

Commentary on Section 8
WORKMANSHIP - ERECTION

As with all other fabrication procedures, erection must be carried out in a logical, expeditious manner, with due regard to safety. A safe method of working is an essential requirement for erection and the NSSS has been compiled to ensure that this is not overlooked.

Between December 1984 and March 1986 the Health and Safety Executive brought out four guidance notes on safe erection of structures and these are cross referenced in both BS 5950 Part 2 and the NSSS. These are Guidance Note GS 28 Parts 1 to 4 *(see Ref. 10)* and they provide guidance on good quality erection procedures. It is not appropriate to highlight just parts of them - they should be read and studied in their entirety. A summary of the contents is as follows:

Part 1 : Initial planning and design.

Pre-site considerations, including design, specification, planning and preparation of proposed methods of work; scheme coordination and management.

Part 2 : Site management and procedures.

Site management, including supervision, coordination and liaison; site preparation, including plant access; stacking, storing and delivering materials; stability, including temporary supports and restraints; holding down and locating arrangements; lifting and handling; advice on interconnecting.

Part 3 : Working places and access.

Access, egress and working heights; minimising the need to work at heights; systems and devices to aid safety at heights.

Part 4 : Legislation and training.

Legislation applicable to erection projects; training.

An important factor in safety is that each person's responsibilities must be clear to all others involved, whether they be designing the structure, planning its erection, managing the erection or actually working on site.

As many more accidents happen when working at height, it is hoped that encouragement will be given to both designers and constructors to adopt pre-assembly at ground level and lifting larger assembled components into place.

The Construction (Design and Management) Regulations 1994

The regulations *(referred to as the C(D&M) Regulations)* came into force on 31 March 1995. They are of course a post-NSSS 3rd. Edition event, but they signal a further step forward in site safety. Under these regulations:

- **Clients** have to appoint a **Planning Supervisor** to coordinate health and safety matters on a project, and to ensure that adequate resources are available in order to comply with the requirements of the regulations.

- **Designers** are required to cooperate with the Planning Supervisor to demonstrate compliance with Health and Safety legislation. Their design must take due account of foreseeable risks to those involved on the construction of a project and its subsequent maintenance. The safety plan prepared must include information about the design of the structure or materials from which it is constructed which might affect health and safety.

- A **Principal Contractor** must take reasonable steps to ensure that all contractors comply with health and safety legislation and with the requirements set out in the safety plan. All such requirements must be in written form and must be brought to the attention of those affected by them.

The terms used in the Regulations do not necessarily coincide with the ones used in the NSSS, or in the Contract Conditions, so care must be taken when referring to them. Thus in the Regulations:

The **Client** means the person for whom a project is carried out. This could be the owner, or a developer, or the agent of the owner or developer. The Client would only be the Employer as defined in NSSS terms, if that person or company had also placed the steelwork contract.

The **Designer** is any person who prepares the design, or who arranges for a person under his control to prepare the design. In many contexts the Designer will be the Architect, but as far as the steel frame is concerned, the Designer is the Engineer as defined in the NSSS.

The **Principal Contractor** is appointed by the Client and can be any contractor who works on the site or who manages construction work on site. If the Steelwork Contractor happens to have a contract where he is responsible for most site activities, it is possible that he would also be appointed the Principal Contractor. However, in many instances, when steelwork is a subcontract, it will be the General Contractor who is appointed to the role.

The person appointed as **Planning Supervisor** can also be appointed as Principal Contractor if he is competent to carry out the function of both appointments; in fact, the Client can appoint himself to these duties if he is competent to perform the functions.

8.1 GENERAL

Erection Method Statement

The **erection method statement** required by **8.1.1** is intended to be prepared by the Steelwork Contractor and submitted to the Engineer before commencing erection work.

An outline erection method statement should have been prepared by the designer *(see GS28/1)* and detailed design will be based upon this. Any aspects that have to be taken into account during the erection stage should be notified to the Steelwork Contractor. *(See 1.5)*.

If the Steelwork Contractor's erection method is based on the designer's proposed erection method, then the method statement should take appropriate account of the aspects notified. The erection method statement should give all details of the erection scheme and procedures to be adopted, stating clearly the drawings, and the latest issue of the drawings, to be used in erection. It should include details of special features and connections as well as the type of fasteners to be used. The stability, when lifted, of individual components and the resistance of single columns to overturning should be critically assessed in conjunction with the design criteria given in 1.3.

In **8.1.2 acceptance of the erection statement** by the Engineer does not alter the Steelwork Contractor's responsibility for the erection scheme, but it does mean that the Engineer must check that the method proposed is valid for his design, has not deviated from the design concept, and comprises a safe method of erection.

The requirements on the designer are given in Regulation 13 of the C(D&M) Regulations "Every *designer has to ensure that any design he prepares ....for the purposes of construction work includes among the design considerations adequate regard .... to the health and safety of any person at work carrying out construction work."* This reinforces the requirements in 8.1.2, which in turn parallels Clause 1 in BS 5950 Part 1. The Engineer should therefore have considered whether safe erection is feasible when preparing his design.

The Engineer should review the whole of the erection method statement to ensure that the Steelwork Contractor demonstrates a safe system of work which accords with the design assumptions made and that all critical matters are dealt with.

Setting out lines and datum levels are provided by the Employer and these are important to the correct positioning and alignment of the structure. **8.1.3** stipulates that they should be adjacent to the work so that sighting and measurements can be made accurately, without the need to transfer from a distance. It is unreasonable to expect the erector to jeopardize the proper alignment of the structure through having to transfer lines and levels from an out of the way corner of the site with several obstacles preventing direct sighting.

The aim of **8.1.4** is that **handling and storage** should be kept to a minimum. It is generally better to arrange transportation with this in mind.

Since it is not always possible to use suitable protection around painted components, it is important that the Steelwork Contractor makes provision for lifting devices on such members to avoid damage to the paint-work. Where stacking is necessary, hardwood timber spacers should be used to avoid damage and abrasion to surfaces. The Steelwork Contractor should ensure that the steelwork is stored in a manner such that it can be lifted in the proper sequence for erection, and that marks are clearly visible at all times.

It is expected that steelwork will be placed in storage areas with provision made to keep it clear of the ground. Components should not be allowed to become covered with mud and dirt on site. Workmen should be discouraged from walking on steelwork which is stacked and awaiting erection.

Containers or bins should be provided for small items and bolt assemblies. It is a stipulation that they are stored under cover in dry conditions. Welding consumables must also be stored in dry conditions since moisture is detrimental to them.

In **8.1.5** provision is made for **damaged steelwork** to be restored "to the standards of manufacture". This is not intended to mean that members cannot be straightened on site. Although straightening can cause localised stressing beyond yield strength of the material, this is not usually detrimental as steel is a malleable material having enough ductility to accept such treatment. However, severe hammering to bring back to shape is not acceptable.

It is necessary on occasions to make some fabrications completely on site. This should be acceptable provided the same standards as applied to shop fabrications are adopted.

Column base packings are often overlooked in the design; **8.1.6** provides some rules for them. 9.4.1 makes acceptable a permitted deviation in concrete foundation level being –30mm, so packs must be available in sufficient thicknesses to accommodate this deviation whatever grout space is provided. The packs should be flat and free from burred edges, and must be positioned to suit column base plate details and the method of plumbing the column.

Although **8.1.7** requires that **grouting of column bases** is not done until a sufficient portion of the structure is aligned, levelled and plumbed, it should not be used as an excuse for delaying grouting for weeks or months on end. Subsequent operations may cause movement in column positions which are not grouted. A column cannot be considered as fully anchored until grouting is complete.

8.2 SITE CONDITIONS

The **Employer's responsibilities** on site, in NSSS terms, are given in **8.2.1;** he may well have other obligations in preparing, maintaining and organising the site (for example, adequate areas for mobile elevated working platforms to operate). Proper hard-standing areas and firm road accesses suitable for the delivery of heavy steel members on standard vehicles is needed.

The site should be free of standing water, and the Employer may have to pump foundation pits clear of water so that work at foundation level can be carried out.

He also has the duty to be aware of positions of all new and existing underground services, and to acquaint the Steelwork Contractor with details of any which are in the working areas being used.

If any overhead obstructions exist which prevent the proper operation of cranes for use in erection, the Employer should arrange temporary or permanent removal. This requirement includes the removal of electric wires and cables which may be a safety hazard during erection or clear marking of routes for cranes and similar vehicles within the site.

In **8.2.2** the Steelwork Contractor's responsibility includes the design of any temporary supports for crane footings. Timber spreaders, or other means should be provided to ensure that the ground bearing capacity under the crane supports is not exceeded. The Steelwork Contractor should be fully aware of the requirements of BS 7121 *Code of practice for safe use of cranes*, which includes including inspection, testing and examination of cranes.

8.3 SAFETY

Safety is re-emphasised once again in this clause. In **8.3.1** reference is made to the advice given in the Health and Safety Executive's Guidance Notes GS 28. Parts 1, 2, 3 & 4 which sets out the **responsibilities of all parties** *(see Ref. 10)*. It is also required that erection is carried out with respects to the appropriate Section of BS 5531 *Code of Practice for safety in erection of structural frames*, which deals with steel, timber, precast concrete and plastic frames. A practical guide to good site practice can also be found in GS 28 Part 2.

The **Steelwork Contractor's responsibilities** for safety are outlined in **8.3.2**, and include compliance with both the Employer's rules for operating the site and current legislation. Much of this is now included in the C(D&M) Regulations and, in particular, in the Safety Plan.

Safety is the subject of much legislation in addition to the C(D&M) Regulations and it is embraced in new European directives. Two publications which contain safety guidance for both office and site based staff are *Structural Steelwork Erection* and *Erector's Manual*, both available from the BCSA (see *Refs. 11 and 12*).

8.4 STABILITY

These clauses are intended to ensure that the structure being erected is fully stable at all times. In this context it means three things:

- the stability of members being lifted and until properly restrained,
- the strength and stability of components to resist temporary erection loading, and
- the overall stability of the structure until permanent features are added.

To avoid instability, temporary restraints may be necessary for any of these categories until all steelwork and other permanent restraints are in position. Short length components are not likely to have stability problems; however, long components and particularly long, deep components may be expected to require restraints when they are being lifted and until some, or all, permanent members are in position.

Particularly vulnerable components are :-

- Lattice girders with single gusset plate joints.

- Narrow flanged beams and plate girders which will have more than one permanent restraint when erected.

- Members to form steel/concrete composite floor beams having to support wet concrete without restraints.

- Members having to resist high wind loads in the temporary condition.

- Members in significant compression.

Special attention is paid in **8.4.1** to **temporary restraints**, which are a design requirement until permanent features are built. This is because it is considered that the Engineer, being the designer, knows which permanent features provide restraints in the finished structure. He is also aware of the design stresses in the components. On the other hand the Steelwork Contractor, when building a complicated frame, may be unaware of these matters. It is therefore incumbent on the Engineer to advise the Steelwork Contractor about these features, particularly where it is not obvious from reading the design drawings *(see C(D&M) Regulations and GS 28)*.

The Steelwork Contractor is required to design and provide any temporary bracing and restraints needed. Full details should be included in the Erection Method Statement.

The purpose of **8.4.2** is to make it clear that the Employer and other subcontractors should not rely on the **temporary restraints** used for erection purposes being kept in position after lining and levelling of the steelwork is complete <u>unless</u> the Steelwork Contractor has been specifically requested to leave the temporary restraints in position. If there is any possibility of the structure distorting before walls and cladding are in position then temporary bracing should obviously remain after steel erection is complete, but the Steelwork Contractor should be informed of the requirement so that it can be properly programmed and costed *(see 1.2(iv) and 1.8)*.

8.5   ERECTION LOADS

Temporary loading from erection equipment should be ascertained and calculations made to ensure that the structure is not overloaded.

The greatest danger is temporary stacking of material from following trades as work proceeds, where point loads can overstress beams as well as causing local damage. The Principal Contractor now has an obligation under C(D&M) Regulations to prevent other contractors from overloading the structure. Stacks of steel decking, concrete units and access scaffolding are common potential causes of overload.

8.6   LINING AND LEVELLING

The purpose of **8.6.1** regarding **alignment** is to ensure a logical sequence to erection activities in lining and levelling, to avoid unnecessary remedial work in having to repeat operations completed earlier.

BS 5964 *Building setting out and measurement* provides much useful information including the transfer of bench marks, whilst BS 7307 *Building Tolerances, Measurement of buildings and building products* deals with the measurement of buildings. It also refers to building tolerances, but does not give detailed requirements for steelwork erection. Section 9 of the NSSS is relevant to permitted deviations.

8.6.2 serves as a reminder that **temperature effects** must be considered in the erection of steelwork. The coefficient for linear expansion of steel is 1.2×10^{-5} per °C, e.g. a 10°C rise in temperature expands 100 metre length of steel by 12mm. Temperature is therefore a factor to take into account when making measurements; both steel tapes and steelwork are affected. In high rise structures which may be partially in shade, measurements are best taken in early morning when temperatures are more constant. Comprehensive information on measuring instruments is given in BS 7334 *Measuring instruments for building construction. (See also notes to 1.3.1 (xiv).)*

8.7   SITE WELDING

The NSSS intends that all welding is of the same quality whether performed on site or in the workshops. Section 5 specifies relevant requirements.

This should not deter the use of site welding since experience has shown that it can be successfully undertaken on both heavy and light steel structures with good overall economy when proper planning for welding has taken place.

Factors to be considered are :-

* floor by floor completion to give good working areas.
* use of light easily erected working platforms.
* protection from inclement weather.
* portable ovens for keeping electrodes moisture free.
* heated quivers for carrying electrodes.
* close attention to steel "carbon equivalent" in accordance with BS 5135 and to steel sulphur content which can contribute to lamellar tearing in conditions where the weld is restrained in tension on cooling.
* carefully designed temporary cleats and brackets to maintain the required gap between components to be welded.
* careful detailing to ensure downhand welding.
* use of details and techniques to avoid the necessity for excessive pre-heating.
* provision of temporary means of support and stability until welding is complete.

8.8 SITE BOLTING

Since a large proportion of all bolting is performed on site, a direct reference to Section 6 is appropriate.

It may not be feasible for an inspection to be made of all bolted joints on site, and it is therefore advisable to keep the procedures for installing and tightening fasteners as simple as possible. The use of standard length fully threaded bolts is one way in which to ensure the correct fastener has been used. *(See notes to 2.8)*.

8.9 CERTIFICATION OF COMPLETION

This clause is designed to ensure that both parties to the contract have signified that the erection work has been satisfactorily completed. The Employer and Steelwork Contractor sign the certificate, prepared by the Steelwork Contractor, when the work is fully complete, or, in the case of phased construction, when the section of the work is complete.

Many other trades follow steelwork erection in order to complete the building. It is therefore in everyone's interest to be satisfied that the steelwork is correct and accepted before further work proceeds. Remedial work at a later date can be very expensive.

The Steelwork Contractor must signify that the work is finished and that he has satisfactorily completed his inspection routines.

The Employer must signify that he has completed any checks he wishes to make and that he accepts the structure, as built, as being satisfactory with regard to the Project Specification.

This clause does not of course indemnify the Steelwork Contractor if any other items of faulty workmanship are found later. However, it may be prudent for both parties to write into the certificate comments about particular topics which have been agreed so that matters are fully recorded if problems arise later.

Commentary on Section 9
WORKMANSHIP - ACCURACY OF ERECTED STEELWORK

Tolerances for erected steelwork are a comparatively recent innovation. British Standard Specifications for steelwork ignored the subject until the advent of BS 5950 in 1985. The limits on the shape of rolled members were always governed by BS 4, and there was an understanding that members would be cut to a length of $\pm 1/16''$ (1.6mm). This one tolerance alone provided adequate safeguards for the erection of steel frames for more than 50 years. Today, only steelwork observes permissible deviations written in specification terms; other materials have generalisations on shape tolerances given in codes of practice, and construction tolerances in BS 5606 *Guide to Accuracy in Building*. A paper by R J Pope provides a current perspective on tolerances in steel construction *(Ref 13)*.

A deviation 'envelope' can be established from the values given in Section 9, in which each member of a structure should be built. Other tolerances are provided, which are probably more important, giving the permissible variation between adjacent members. When considering the fixing of floors, walls, cladding and glazing to steelwork, it is the relative accuracy of adjacent members which is often most critical, not an exact position in space.

It is not difficult to show that if the cumulative effect of the fabrication and material deviations coupled with erection deviations are considered, some members may be outside the deviation 'envelope' required for fitting other components directly to the steel. This is why, in the preparation of drawings, the fabrication deviations have to be recognised and provision made for adjustment where necessary to comply with the permissible erection deviations. *(See 3.4.6.)*

It would be possible to write endlessly on the subject of permissible deviations in trying to cover every conceivable situation. However, it is considered that Section 9, coupled with the fabrication tolerances in Section 7, give a basis for use in most structures. Whilst the Steelwork Contractor must make every effort to maintain the standards laid down, there will be occasions when the question 'Does it matter?' will arise. Common sense should prevail. Stability of the erected frame must always be observed, but some judgement may be exercised in issues relating to exact geometry of the frame. There is little to gain in holding up a contract while the Steelwork Contractor, who has safely and adequately erected the steelwork, has to push and pull the steel members again to achieve the last few millimetres in position of the column or beam.

9.1   FOUNDATIONS

The NSSS does not have specific tolerances on the position of the concrete foundations other than levels since it was considered that the obligation to have the holding down bolts correctly positioned would ensure an acceptable location for the steel base plate. *(See 9.4.3 and 9.4.4)*

Experience dictates the wisdom of requiring the Steelwork Contractor to inspect the foundations before erection commences. There is little point in bringing an erection team to the site when remedial work is needed to the levels of foundations or the position of H.D. bolts. It has to be remembered that the Steelwork Contractor is inspecting for foundation bolt position only. Acceptance of the foundations by the Steelwork Contractor implies nothing more than that the holding down bolts are dimensionally within the permitted deviations relative to a setting out point and building lines provided by others. It does not, in any way, relieve the foundation contractor of the obligation he should have in his own contract regarding the true position of the foundations.

9.2   STEELWORK

The permitted deviations specified in the NSSS were determined after discussing with Engineers what could reasonably be accepted, and with Steelwork Contractors, as to what was reasonably practical. A higher degree of accuracy would be reflected in a much higher cost. Deviations given in this Section are for the bare frame, acting under self weight only. When other loading is encountered at the time of measurement, the Engineer should advise the predicted change in the geometry of the frame.

This clause makes the point that temperature must be considered and a cross reference is made to 8.6.2 which deals with the effect of temperature on dimensions. Where there is a problem, it is best to take dimensions early in the day when the whole structure is more likely to be at an even temperature.

There are other matters which affect plumbing and alignment, and which are sometimes wrongly construed as being caused by out of tolerance members.

They are:

- member or frame deflections under load have not been fully recognised.
- temperature effects such that expansion joints should have been included.
- members lacking torsional stiffness allowing other elements to rotate.
- the effect of asymmetrical loading has been overlooked.

Most of the diagrams shown in Section 9 indicate rolled sections. The intention of the NSSS is that RHS, CHS and larger built-up shapes are to be erected to the same tolerances as those permitted in rolled shapes portrayed in the diagrams.

Deviations in this Section are shown to be at the ends of a members or at node points where the level and position can be adjusted.

9.3 INFORMATION FOR OTHER CONTRACTORS

It is most important that other contractors associated with a project, know the permitted deviations in fabrication and erection and also the variation that they may find in rolled steel sections *(see Fig. 2.3)*. The Engineer has to advise the contractors following steel erection of the permitted deviations in steelwork dimensions, so that clearances are allowed and the connections to the steel frame are made to suit and have the appropriate range of adjustment *(see notes to 7.3)*.

The importance of having adequate adjustment in cladding attachments cannot be over-emphasized, particularly when large rigid panels have to be attached to more than one member. Having to cut and re-weld brackets at site can be very expensive and plays havoc with the erection programme.

Similarly, the Engineer has to advise the Foundation Contractor of the degree of accuracy required in positioning foundation bolts *(see 9.4)*. The Steelwork Contractor will detail and fabricate column bases in the expectation of these values being achieved.

9.4 PERMITTED DEVIATIONS FOR FOUNDATIONS, WALLS AND FOUNDATION BOLTS

Foundations and column bases

The permitted deviations affecting a column base plate are shown together in Figure 9.1. It can be seen that they have been arranged so that the tolerance on the concrete foundation permits a lower level but not a higher one; to compensate, the setting out of foundation bolts is required to be higher

than the theoretical level. Practical use can then be made of base packs
(see 8.1.5) to level the column so that the components connected to it are
within a level of ±10 mm as required by 9.5.7.

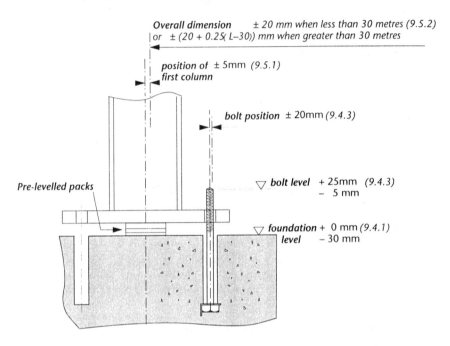

Overall dimension   ± 20 mm when less than 30 metres (9.5.2)
or   ± (20 + 0.25(L–30)) mm when greater than 30 metres

position of ± 5mm (9.5.1)
first column

bolt position ± 20mm (9.4.3)

Pre-levelled packs

▽ *bolt level   + 25mm (9.4.3)*
          – 5 mm

▽ *foundation + 0 mm (9.4.1)*
 level   – 30 mm

Figure 9.1 Tolerances for adjustable column bases

It should be noted that the deviations in **9.4.3** are referring to a single
bolt or a bolt group. The understanding is that a group of bolts will be set
using a template to ensure the correct relationship of all bolts in the group
for a column base plate.

Whilst the majority of Steelwork Contractors prefer adjustable foundation
bolts, some have an undoubted preference for preset, non-adjustable, bolts.
When this type of foundation bolt can be correctly located, in plan, to within
3 mm, as required by **9.4.4**, it has the distinct advantage of automatically
aligning the columns, and adjustment being confined to levelling and
plumbing. It has obvious advantages for portal frames where the natural
tendency for the frame to spread is prevented once the base is located on the
bolts.

Vertical walls

The Engineer dealing with a steelwork connection to a vertical wall face has to make allowances for a wall being up to 25mm out of position in accordance with **9.4.2**. Pre-set bolts must of course be able to cover the 50mm 'envelope' on the position of the walls. In **9.4.4** provision is made for this with 45mm being added to the normal protrusion needed to accommodate the nut and washers plus a further tolerance. Figure 9.2 shows the type of connection envisaged.

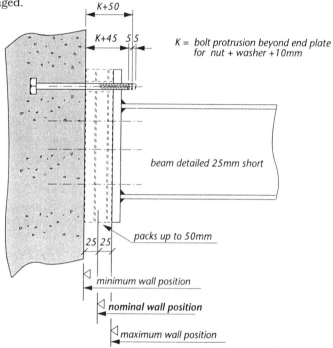

K = bolt protrusion beyond end plate for nut + washer +10mm

Figure 9.2 Connection to a wall

If the connection detail shown is developing vertical shear forces it would be necessary to design for bolt bending to take account of load eccentricity, unless the packs shown were welded to the beam end plate.

9.5 PERMITTED DEVIATIONS OF ERECTED COMPONENTS

In the NSSS emphasis is given to the location of the first column erected. The philosophy is that it should be located within 5mm of its true position **(9.5.1)**; the control on fabrication deviations should automatically ensure that other column bases are in an acceptable position when the frame is plumbed and aligned. However, some insurance for this is provided by the overall permissible plan dimensions in **9.5.2**, combined with a method of adjustment being included in the detailing (e.g. beams generally detailed one millimetre short, with occasional shim plates).

The Foundation Contractor must be made aware that the requirements of **9.5.2** also apply to the position of the foundation for the most remote column, relative to the foundation for the column first erected. This is essential if the foundation bolts are to be correctly placed in the remote foundation. The Steelwork Contractor must then ensure that the steelwork has sufficient adjustment so that the overall length is maintained within the tolerances stipulated in **9.5.2**.

Whilst **9.5.3** requires **single storey columns** being plumb to within 1/600 of the height, portal frame columns are excluded. Nevertheless, if the Engineer specifies the amount of pre-set required, in accordance with 1.3.1 (xiv), the specified deviations can be used as a tolerance on the pre-set dimension.

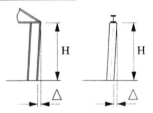

$\triangle = \pm$ H/600 or 5mm
whichever is greater
Max $= \pm$ 25mm

NSSS 9.5.3
Column plumb: single-storey

The **column plumb limits for multi-storey** buildings in **9.5.4** were considered particularly applicable to medium height structures of five to fifteen storeys. However, the maximum permitted deviation of 50mm may be equally suitable for taller structures.

Columns in the area of lift shafts may need more stringent limits. The elevator guide tolerances commonly restrict the plumb of adjacent columns, but the design of the frame may also have to restrict the horizontal displacement under wind forces. Therefore deflection and plumb may have to be considered together. *(See SCI Publication P103 Ref. 14)*

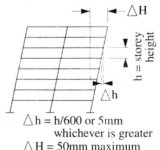

$\triangle h = $ h/600 or 5mm
whichever is greater
$\triangle H = $ 50mm maximum

NSSS 9.5.4
Column plumb: multi-storey

Final plumbing of columns is likely to determine the **gap between bearing surfaces** at splice positions. In **9.5.5** the permissible gap is given. As further load is applied to elements in bearing, the parts in contact will yield until there is sufficient area to carry the load, and the gap closes. However, if the gap is excessive, unwanted eccentricity may be introduced. Experience shows that the permitted deviation is satisfactory for most structural conditions and this is supported by test results *(Ref. 15.)*

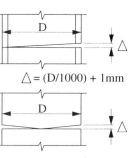

$\triangle = (D/1000) + 1\text{mm}$

NSSS 9.5.5
Gap: bearing surfaces

If the deviation in alignment of adjacent columns takes place in an abrupt way, all manner of complications could arise and make life impossible for other trades making connections to the steel frame. 9.5.4, by itself, could theoretically allow two columns to vary with an out of plumb dimension at the top of 100 mm, if they were each at the maximum deviation in opposite directions; this is clearly absurd. The alignment of **adjacent perimeter columns** is therefore introduced in **9.5.6** to avoid this anomaly. The intention is that adjacent perimeter columns should vary by no more than the deviation indicated at the base or splice. These positions are stipulated because they are the points where some adjustment is possible. A varying degree of straightness in adjacent columns may cause a greater discrepancy between these points.

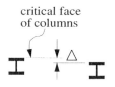

critical face
of columns

$\triangle = 10$ mm

NSSS 9.5.6
Column alignment

The three clauses **9.5.7**, **9.5.8** and **9.5.9** are intended to ensure a maximum deviation of ±10 mm for **floor beams** at the connection-to-column positions, but without any abrupt changes of level. This is ensured by the 5 mm rule applying to the opposite ends of a beam or for the ends of adjacent beams.

Since the straightness and deflection of a beam will vary along its length, the actual variation in level at intermediate points could theoretically exceed the limits given at connection-to-column positions. In practice, deviations which occur in normal situations should not cause stability problems or concern.

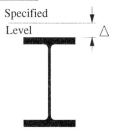

Specified
Level

$\triangle = \pm 10\text{mm}$

NSSS 9.5.7
Floor beam level

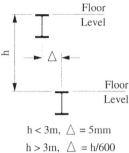

The **alignment vertically of beams** is largely related to the plumb of the columns. Clause **9.5.10** therefore allows the same deviation as in 9.5.4 for column alignment.

$h < 3m, \triangle = 5mm$

$h > 3m, \triangle = h/600$

NSSS 9.5.10
Beam alignment

Crane gantries have a particular need for accuracy in erection if the crane is to operate freely with minimal maintenance. Clauses **9.5.11**, **9.5.12** and **9.5.13** were formulated to achieve this.

When the gantry is at a high level the permitted deviation from plumb of the column at crane girder level is such that, in practice, adequate adjustment is almost always provided in the connection of the crane girder to the column.

If the columns are erected to the accuracy demanded in **9.5.11**, and proper adjustment allowed, then it should be possible to align each crane track to ±5mm of its nominal position.

An alignment limit between adjacent columns is not stated, but it should not be allowed to be more than that provided by 9.5.6; i.e. an absolute discrepancy of 10 mm for any height of column.

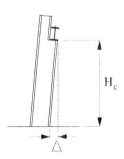

$\triangle = \pm H_c / 1000$ or 5mm
whichever is greater

$Max = \pm 25mm$

NSSS 9.5.11
Crane gantry: plumb

The overall allowance of ±10mm in **9.5.12** takes account of the possibility of each track being at its maximum deviation in opposite directions. It is prudent to connect the crane rails to the crane girder using adjustable clips if the more onerous tolerances required by Appendix F of BS 466 *Specification for power driven overhead travelling cranes* are to be achieved.

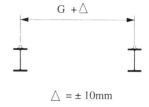

$\triangle = \pm 10$mm

NSSS 9.5.12
Crane gantry: gauge

The intention of **9.5.13** is that adjacent crane rails should be, to all intents and purposes, level. A deviation of 0.5mm is hardly measurable and is simply a definition of 'levelness'. As the rail joints are unlikely to coincide with the crane girder joints, it may be necessary to grind the top surface at, say a slope of 1:100, to ensure that adjacent rails are 'level' at the joint.

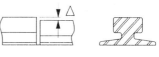

$\triangle = 0.5$mm

NSSS 9.5.13
Crane rail joint

Commentary on Section 10
PROTECTIVE TREATMENT

The aim of Section 10 is to provide the ground rules for good practice in the preparation for surface protective treatment of structural steelwork and the application of the protective treatments. The actual treatment to be used should be stated in the Project Specification as discussed in the notes to Section 1. *(See 1.6.)*

Although outside the scope of these notes, it should be recognized that designers and detailers of steelwork should endeavour to minimise the possibility of corrosion by avoiding traps for moisture and ensuring easy access to surfaces requiring to be painted. Detailed notes on designing to prevent corrosion can be found in BS 5493 *Code of practice for Protective coating of iron and steel structures against corrosion, Appendix A.*

Many steel building structures, where internal steelwork is situated in a dry environment, do not require protective treatment. However, where it is required it should be specified in a manner which is clear, concise and technically achievable. A good protective treatment specification does not give scope for controversy and dispute; it also enables the planning of the work so as to provide conditions necessary for successful protection. Section 10 provides rules to achieve this.

It should be remembered that mistakes made in the type or application of protective treatments can prove very expensive to rectify. Remedial work may have to be carried out in very difficult locations if the mistake is found after the steelwork has been erected.

10.1 GENERAL

The importance of a correctly prepared Protective Treatment Specification cannot be over-emphasised. Detailed guidance notes can be found in Section 3 of BS 5493. The current edition is dated 1977 with amendments issued in 1984 and 1993. Paint materials have been further developed since then and 'paint life' has changed, so paint manufacturers should be approached for latest guidance on these matters. Unless experienced in the preparation of protective treatment specifications, expert advice should be sought. Steelwork Contractors, Paint Manufactures and British Steel's advisory services will advise suitable treatments for particular use.

Of course, as well as providing protection against corrosion, treatments may also give fire protection when intumescent paints are used, or simply be provided for cosmetic reasons.

The value of the Steelwork Contractor providing a written **Method Statement**, as required by **10.1.2**, is that this demonstrates to the Employer that the protective treatment specification and the quality control and inspection procedures required to ensure conformity with the specification have been satisfactorily described and are understood. A good method statement, like the specification itself, can serve as a basis for early clarification of requirements and so reduce the likelihood of disputes and misunderstandings once the work is in progress.

10.1.3, dealing with **working conditions**, refers to Clause 23 of BS 5493: *Code of Practice for Protective coating of iron and steel against corrosion*. It gives full guidance on the conditions under which protective coatings should be applied and includes the following, which should be taken into account when preparing the Method Statement:

- The environmental conditions in which coating work can proceed to suit the application properties of the materials being used, and suggested practical methods for improving the immediate environment of the application and drying and curing processes.

- Some materials and application processes are not so sensitive to inclement conditions and some relaxation may be permitted in the Specification.

- The temperatures of the environment and of the surface to be coated can affect the following characteristics of paint before, during and after application:
  (a) Solvent retention.
  (b) Viscosity of liquid coatings and consequently the brushing and spraying properties.
  (c) Thickness and appearance of dry films.
  (d) Drying time.
  (e) Pot-life, curing time and overcoating periods of two-pack materials.

- The Specification should stipulate that coatings are not applied to surfaces where the relative humidity of the atmosphere is such that:

  (a)   Condensation is present on the surface; or

  (b)   it will affect the application and/or drying of the coating.

- When heating is being used to control the relative humidity of the environment in enclosed working spaces, it is usual to specify that heaters which exhaust combustion gases into the working environment should not be used.

- Clauses may be included in a Specification which will permit preliminary preparation to be done in the open under conditions which would not be suitable for final preparation.

- Precautions should be taken to ensure that prepared surfaces or surfaces with a wet coating are not contaminated.

- The entire programme of surface preparation and coating application should be planned in detail at an early stage.

- Adequate lighting of surfaces is essential with sufficient illumination for work and inspection.

- Coating work carried out must comply with all relevant Health and Safety Regulations (They are summarised in Section six of BS 5493 but a check should be made to ensure all current statutory requirements are met).

BS 5493 provides a more comprehensive discussion on these matters. There is no doubt that the use of environmentally controlled shop conditions affords the best means of realising the full potential of protective systems. Such shop conditions are always preferable, and with some specifications, essential.

The **storage of materials** for protective treatment should be part of the Specification and proper use of them, in sequential order of delivery, should be ensured by a batch date-marking system. Most paint materials have a shelf life which is measured in months rather than years and many are susceptible to frost. **10.1.4** requires that the manufacturer's recommendations regarding storage and shelf life are adhered to. The shelf life is the time that the material will remain in good condition when stored under normal conditions and applies only to materials stored unopened in their original containers (BS 2015 - *Glossary of Paint Terms*, refers to this).

Handling and storage procedures to minimize damage are covered by **10.1.6**, but specific requirements cannot be given since structural steel components vary in size and shape. The aim is for minimum re-handling of steelwork and a storage system should be arranged to suit this. Where painted steelwork can be stacked, it should be separated with timber spacers. Care must be taken to ensure that the lifting arrangements do not allow the chain slings to slip if anything is placed around the steelwork to protect the paint or metallic coating.

Some protective systems are more robust than others and capable of withstanding rougher treatment but it is virtually impossible to avoid some damage to paint treatments in handling and transport. For this reason it is strongly recommended that final paint coats are applied at site, after erection.

10.2 SURFACE PREPARATION

It is not intended in **10.2.1**, that all mill-scale and rust be removed by **wire brushing**; that is impossible. Unless freshly delivered, under cover, from the rolling mill, or from an enclosed stockyard, the steel surface will be a variable mix of mill-scale and rust as a result of weathering or moisture condensation over prolonged periods. Wire brushing should only be specified to remove loose mill-scale followed by cleaning to remove dust, oil and grease and is generally only used on steel which remains in a dry environment.

Blast Cleaning in 10.2.2 now adopts BS 7079 Part A1 in place of the Swedish specification previously used. The pictorial illustrations of various levels of blast cleaned surfaces are the same as the Swedish SIS 05 59 00. Good workmanship in blast cleaning is the subject of the rest of this clause.

Adherence of paint or metallic coatings depends upon the roughness and profile of the surface. Shot and well used cut wire abrasives give a smoother blast profile than that produced by angular abrasives. However, the use of the latter is preferred for some paint coatings and is essential for metal sprayed coatings. If in doubt about the surface to specify, advice from the paint manufacturers should be sought. Metallic abrasives are designated as Group E in BS 7079, whilst non-metallic abrasives are Group F.

The reference to cleanliness is intended to include freedom from dampness, since the blast cleaned surface will rapidly deteriorate where moisture can collect. In this respect, night time temperatures should be taken at the prevailing humidity levels within the fabrication shop to check if condensation may form on the steel surface. The steel will have a thermal lag compared with the surrounding air and if shop heating is discontinuous, steel that has cooled overnight and then exposed to warmed air at the start of the day shift may suffer condensation, especially between autumn and spring.

One of the consequences of effective blast cleaning may be to expose a defect which is present on the steel surface. 10.2.3 requires that the defect is dealt with in accordance with 2.5 (ii) or (iii), but steel surfaces are generally more easily rectified at an at an early stage before fabrication begins.

10.3 SPRAYED METAL COATINGS

Sprayed metal coatings, sometimes used as the only protection, and sometimes with further paint coats added, can provide a long life. After spraying the surface is usually given a sealing treatment. If this is not done it may suffer from white staining if stored in damp and un-ventilated conditions. (Further comments about white rust staining will be found in 10.4.)

Extensively **damaged areas**, defined as being over $10cm^2$, are required by **10.3.3** to be repaired after cleaning and washing down. If a needle gun is used in the cleaning, care should be taken not to create a weakness in the area around the damage.

In **10.3.4**, smaller damaged areas on both aluminium and zinc sprayed steel are permitted to be repaired by a suitable paint treatment. The use of two-pack zinc-rich epoxy paints is the preferred method.

It is essential that in cleaning the damaged area, the surface roughness is controlled and the paint coating thickness is sufficient to adequately cover any peaks on this surface. Zinc rich primers can also suffer from white staining if stored in damp and un-ventilated conditions.

10.4 GALVANIZING

Hot dip galvanizing by immersing suitably cleaned steel in a bath of molten zinc after fabrication is another long life process. BS 729 deals with the composition of zinc in the galvanizing bath, appearance and uniformity of coating, distortion, cracking and embrittlement of the base metal of the item being galvanized.

The NSSS does not specify any particular rules for care after galvanizing, but it is sometimes prone to a white storage stain (white rust) if it is stored in damp un-ventilated conditions. In extreme cases, the stain is voluminous and unsightly. Unless white rust is removed, any intended painting may suffer from adhesion problems. Light deposits of white rust can usually be removed with a stiff bristle brush but heavier deposits may need to be chemically removed by treating with a 5% solution of sodium or potassium dichromate with the addition of 0.5% by volume of concentrated sulphuric acid. The solution is applied by brush and left to react for about half a minute before thoroughly rinsing off and drying. Some loss of galvanized coating thickness will have occurred over the white rust areas and the metal thickness should be checked with a suitable coating thickness gauge.

Vent holes must always be provided in hollow sections which are to be galvanized; an explosion can occur without them. If the Engineer specifies it, such holes may be sealed after galvanizing with a suitable plug.

Galvanized steel is not readily damaged but, as with all protective coatings, lifting and handling procedures should be specified to minimise coating damage. (See 10.1.)

10.5 PAINT TREATMENTS

Matters considered as important for good practice in painting are covered in this clause, but it is not intended to detract from the use of BS 5493. The British Standard classifies recommended methods of surface protection of structures exposed to environments commonly encountered. It describes the various methods in detail and gives guidance on how to specify a chosen protective system, how to ensure its correct application, and how it should be maintained.

Guides are also available recommending suitable paint treatments for steelwork used in a variety of situations. British Steel and BCSA provide a series of such guides dealing with steelwork in interior and exterior environments and in cavity walls and the background to corrosion protection *(see Ref. 16)*. Note must also be taken of any specialist supplier's advice.

The nature of the **surface** to which coatings are applied will always influence their performance. Hence **10.5.1** requires clean dry surfaces, free from loose mill scale. The probability of failure increases with the amount of mill scale adhering to the surfaces, so blast cleaning must always be the preferred treatment.

10.5.2 does not call for treatment to **surfaces embedded in concrete**, since concrete inhibits corrosion if the provisions of BS 8110 are followed. Generally the minimum depth of cover required to ensure this may vary from 20mm for internal use to over 60mm for the most severe conditions.

The **'pot life'** referred to in **10.5.3** is the period after mixing the components of two-pack paints during which time the paint remains usable. This usable period will be stated on the paint manufacturer's technical data sheet for the product. Pot life is not a term which is generally applicable to other paints and it must be carefully distinguished from shelf life which is mentioned in 10.1.4.

An additional **stripe coat** of primer or undercoat required in **10.5.5** may be considered a nuisance to some. Nevertheless it is there to ensure that all surfaces at critical positions have adequate covering (refer to BS 5493 clause 18 for guidance.

10.5.7 and **10.5.8** are provided to take account of the factors which concern specifiers of painting on site. More information on working conditions when applying protective coating can be obtained from BS 5493 Section 23.

As with the application of stripe coats, the aim of these two clauses is to ensure that full protection is given after site bolting and welding. After cleaning, all parts of assemblies must have a touch-in treatment equal to the surrounding surface.

An exception may be made for bolted assemblies which have their own protective treatment, but the requirements of 10.5.8 must be followed.

Commentary on Section 11
QUALITY ASSURANCE

The Steel Construction Industry deals with a product which must be tailored to suit each project with regular differences in size, quantity and complexity; the nature of the work is such that repetition of large numbers of identical components is a rarity. Hence, full commitment has to be given to Quality Assurance, as a means to satisfy the requirements of a contract, by completing the project to an acceptable, consistent standard of workmanship, safely, on time, and at an agreed price.

This section has been included in the NSSS to ensure that the Steelwork Contractor has good management procedures, applied company wide, whether or not the company is accredited under a formal scheme. In fact, the whole of the NSSS is a response to the need to demonstrate compliance with the contract specification.

The benefits of having a formalised **management system**, as required by **11.1**, have become increasingly recognised since the introduction of QA precepts in the early 1980s. The benefit to the Employer is that the Works are supported by accurate records, the need for external inspection and verification is reduced, and the Employer's costs in supervising the steelwork contract can be reduced. The benefit to the Steelwork Contractor is that areas of potential misunderstanding and uncertain responsibilities are avoided, with a reduced level of error.

11.1 QUALITY SYSTEM

The quality system should demonstrate the strength and competence of the Steelwork Contractor's resources, and should address management and facilities, equipment and skills, processes and procedures.

Those procedures which should be covered in a **Quality system** (as a minimum) are listed in **11.1.2**. Many of them are the subject of a section in the NSSS, in which case they are not discussed again here but references are included to guide the reader to the appropriate section.

In order to meet the requirement of **11.1.3 (i)** it is considered that the Steelwork Contractor should have properly documented procedures to indicate the objectives, responsibilities, interfaces and performance of each part of his operations; accredited companies, dealt with in **11.1.3 (ii)**, will have them as a matter of course.

(i)   Project management and planning

This comprises procedures to ensure the correct interpretation of the Employer's specified requirements (see Section 1), and to establish and implement contract control arrangements. It includes:

*   Delegated functions by subcontractors and others.

*   A contract management organisation plan which names key personnel, their function in the contract and lines of communication.

*   Appropriate planning procedures throughout the contract, and monitoring of performance and progress.

*   A quality plan which outlines the system for the contract and the contract programme.

*   Appropriate input to the Health and Safety Plan to comply with the C(D&M) Regulations (see notes in Section 8).

*   Appropriate consideration of Environmental Protection.

*   A method for controlling variations, changes and concessions which take place during the contract.

*   Contract management review procedures and method for implementation of remedial actions.

(ii)   Design control

The various levels of design responsibility within a contract are discussed in detail in Section 1. The Steelwork Contractor should have established procedures to control and verify the contract requirements for design. These may include:

*   A design plan defining the principal design activities in a logical sequence, the type of design output and target dates to meet the programme requirements and allocation of design responsibilities.

*   Design of the structure so it can be safely erected, bearing in mind that the Engineer preparing the design must take account of safety and stability aspects of the Erection Method Statement (see 8.1.1).

- Design documentation, production and checking procedures (verification).
- A check that Software used in the design has been validated.
- Procedures for the acceptance of drawings by the Engineer (see 3.6).
- Procedures for dealing with variations.
- Handling and transportation requirements for unusually shaped or large components to ensure stability during movement.
- A formal documented review of the design before issue for detail drawing.

(iii) Documentation control

Procedures should be established to control all contract documentation both to and from the Steelwork Contractor. The system should ensure that all documents issued are identified with the current revision status and prevents the use of invalid or obsolete documents in-house or by subcontractors.

(iv) Material purchasing

Materials and subcontracted services should be purchased from sources which have the appropriate capability in terms of facilities, skills, experience and preferably, product quality and procedures complying with the requirement of BS EN ISO 9000. Purchase orders should clearly identify:

- Technical specification.
- Requirement for inspection and tests.
- Quantity, delivery.
- Protection and packing
- Certification and traceability *(see notes to 2.3)*.

(v) Detail drawing preparation

The essential references for fabrication and erection of structural steelwork are the drawings covered by Section 3 of the NSSS. Drawing office procedures should ensure that the Employer's and Engineer's requirements are correctly interpreted. The following should be noted:

- All drawings should be checked *(see notes to 3.4)*.
- Fabrication drawings should identify all information necessary for fabrication to proceed *(see notes to 3.5)*.
- Drawings are normally expected to refer to the protective coating requirements.
- Although not common practice, it is prudent to have a standard note on all drawings stating that fabrication and erection tolerances are those given in Sections 7 and 9 respectively of the NSSS.

- Approval responsibilities for drawings should be fully understood by the Steelwork Contractor and the Engineer *(see notes to 3.6)*.

- Controls should be in place to ensure that when changes are made to drawings, the new drawings are issued and out-of-date prints are retrieved.

(vi) Fabrication

Workmanship for all aspects of fabrication is dealt with in Sections 4, 5, 6 and 7 of the NSSS. The Steelwork Contractor should be able to demonstrate documented plans exist in the following areas:

- Maintenance and calibration of all process equipment.

- Production Planning.

- Process and Production Control.

(vii) Inspection and testing

Inspection of fabricated steelwork components is covered by 4.9 in the NSSS whilst non-destructive testing and inspection of welds is the subject of 5.5, with Tables 1 and 2 dealing with the scope of inspection, acceptance criteria and corrective action for welds. A quality plan should define:

- Noncompliance and concession reporting procedures.

- Corrective action repair procedures.

- Release/rejection procedures.

- Inspection and test certification.

(viii) Surface preparation and protective treatment

Surface preparation and protective treatment is covered by Section 10. Procedures should be established for all these matters and additionally, for:

- pre-process inspection of fabricated items.

- Curing time between coats and prior to despatch.

- Inspection of components after treatment.

- Corrective action and repairs in the workshops and on site.

- Identification of painted items

(ix) Erection

Erection of steelwork is covered by Sections 8 and 9. Procedures should be established for all these matters and additionally, for:

- Pre-erection survey and formal acceptance of foundations.

- Phased component delivery.

- Safety and environmental controls.

- Operative certification.

- Plant and equipment certification.

- Inspection of each phase of construction.

(xi)   Health and safety

Legislation appropriate to Health and Safety in the steelwork industry is discussed in the Introduction. In addition, the need for a Safety Plan under the C(D&M) Regulations is listed under (i) above and in Section 8. Procedures must include, among other aspects, items such as:

- Identification and control of substances hazardous to health including safe use and disposal of each substance.

- The use of personal protective equipment and clothing by all employees engaged on the contract, and by authorised visitors.

- Information and training to be given on the correct use of personal protective equipment for specific processes and operations.

(xii)   Records

Records may be expected to include, but not be limited to:

- Drawings and design calculations and documentation registers

- Certificates for materials and consumables

- Calibration of equipment

- Weld procedures, concessions etc.

- Inspection and test reports

- Delivery schedules

- Surveys and final inspection results

- Completion of erection and handover certification.

System acceptance

If the Steelwork Contractor is in a third party Quality Assurance Certification Scheme, then the procedures described above will be part of his everyday work, and **System Acceptance** in terms of **11.1.3** should be automatic. However, if the Steelwork Contractor is not in such a scheme, the Employer may wish to make an audit to satisfy himself that the system of management is of a comparable standard to that of an accredited company. The records of previous work, sample documentation and qualifications of the workforce may be examined.

It is not obligatory for Steelwork Contractors to be in a Third Party Q. A. Certification Scheme, however, the position is changing and it is becoming a requirement of many clients that the Steelwork Contractor be approved under a scheme which is accepted as being in accordance with BS/EN/ISO 9000:1994, which has now replaced BS 5750/EN 29000 noted in the NSSS. If accreditation is to include design, then it must be under Part 1 of the document. If design is not included accreditation is to Part 2.

Quality Assurance Certification Schemes in the United Kingdom are certified by the Association of Certification Bodies. Q.A. Schemes include Lloyds Register and BSI for a range of industries, but others specialise in a particular industry or product range.

The Steel Construction Scheme Ltd. (SCQA) is the acknowledged expert in assessing and certifying Steelwork Contractors and their associated products. The Scheme has a requirement for internal quality audits to be carried out at regular intervals.

11.2 ADDITIONAL INSPECTION AND TESTS

In addition to the mandatory tests and inspections required by the NSSS and those provided in the quality system, the Employer can write into the Project Specification a requirement for a particular test or inspection he wishes to witness, or points where he wishes a nominated third party to carry out an inspection. *(See notes on 1.4(vi).)*

Where a test or inspection is to be witnessed and the period of advance notice has been stated, the Steelwork Contractor is obliged to provide the necessary facilities for the Employer's inspector.

11.3 RECORDS

The NSSS envisages that during the period of the contract the Employer may wish to see records, which have to be made available.

All inspection and test records should be retained for a minimum period of five years or longer if required by the contract.

REFERENCES

(1) *National Structural Steelwork Specification for Building Construction*
3RD. Edition - Publication No.203/94
British Constructional Steelwork Association Ltd
The Steel Construction Institute.

(2) *Joints in Simple Construction*
Volume 1: "Design Methods" (2nd Edition) Publication No.P205/93
Volume 2: "Practical Applications" Publication No.P206/92
The Steel Construction Institute
British Constructional Steelwork Association Ltd.

(3) *Design Guide for Wind Loads on Unclad Framed Building Structures*
M.R. Willford and A.C. Allsop.
Building Structures During Construction, (Report BR173)
Building Research Establishment 1990.

(4) *Steel Detailers Manual*
A Haywood and F Weare
Publication No.EP15
Blackwell Science Ltd.

(5) *Introduction to Welding of Structural Steelwork*
J L Pratt
Publication No.P014/89
The Steel Construction Institute

(6) *Weld Quality Levels*
M H Ogle
National Steelwork Conference 1986
British Constructional Steelwork Association Ltd.

(7) *Joints in Steel Construction - Moment Connections*
Publication No.P207/95
The Steel Construction Institute
British Constructional Steelwork Association Ltd.

(8) *Use of fully threaded bolts for connections in structural steelwork*
for buildings
G W Owens
Journal of the Institution of Structural Engineers
1st September 1992 pp 297-300

(9) *Lack of fit in steel structures*
A P Mann and L J Morris
Construction Industries Research And Information Association
Report 87

(10) *Guidance Note GS 28   Safe erection of structures*
      Part 1: initial planning and design (1984)
      Part 2: site management and procedures (1985)
      Part 3: working places and access (1986)
      Part 4: Legislation and Training (1986)
      Health and Safety Executive
      Her Majesty's Stationary Office

(11) *Structural Steelwork Erection*
      W H Arch
      Publication No.20/89
      British Constructional Steelwork Association Ltd.

(12) *Erector's Manual*
      E G Lovejoy, H R Stamper, P H Allen
      2nd Edition
      Publication No.16/93
      British Constructional Steelwork Association Ltd.

(13) *Tolerances in Steel Construction*
      Roger Pope
      New Steel Construction
      April 1995
      The Steel Construction Institute
      British Constructional Steelwork Association Ltd.

(14) *Interfaces: Electric lift installations in steel framed buildings*
      R G Ogden
      Publication No.P103
      The Steel Construction Institute
      National association of Lift Makers

(15) *Capacity of Columns with Splice Imperfections*
      A P Popov and R M Stephen
      Engineering Journal
      American Institute of Steel Construction
      1st Quarter 1977 pp 16-23

(16) *Prevention of Corrosion of Structural Steels &*
      Corrosion Protection leaflets:
      • *Building Interiors*
      • *Building Exteriors*
      • *Perimeter Walls*
      • *Building Refurbishment*
      • *Indoor Swimming Pools*
      Publications Department British Steel plc
      Steel House REDCAR TS10 5QW

New European (BS EN) Standards

issued or pending since the 3rd Edition
of the NSSS was published in July 1994, as at June 1996

Quoted Standard	New Standard	Subject
BS 5950: Part 2	ENV 1090-1-1	Workmanship on steel structures
BS 5135	pr EN 1011	Welding
	BS EN 719	Welding Co-ordination
	BS EN 29692	Joint preparation for metal arc welding
	BS EN 439	Shield gas for arc welding and cutting
	pr EN 1289	NDT - DPI of welds - acceptance levels
	pr EN 1290	NDT - MPI of welds - method
	pr EN 1291	NDT - MPI of welds - acceptance levels
BS 639	BS EN 499 arc welding	Covered electrodes for manual metal
BS 5493	ISO/DIS 12944	Protection of Steelwork - Paint Coatings
	pr EN ISO14713	Protection of Steelwork - Metallic Coatings
BS 729	pr EN ISO 1461	Hot Dipped Galvanised Coatings
BS 4848: Part 2	pr EN ISO 10056	Equal and Unequal Angles - Dimensions
BS 4190	BS EN 24016	Bolts - Grade 4.6
	BS EN 24034	Nuts - Grade 4.6
BS 3692	BS EN 24014	Bolts - Grade 8.8
	BS EN 24032	Nuts - Grade 8.8
BS 5750/EN 29000	BS EN ISO 9000	Quality management and quality assurance

British Constructional Steelwork Association

Members

Allott Bros & Leigh Ltd	01709 364115	Hawkes Construction Co	01708 725538
The Angle Ring Co Ltd	0121 557 7241	Hawkins Structures Ltd	01723 584121
Archbell Greenwood Structures Ltd	01253 778855	Horwich Steelworks Ltd	01204 695989
Arnold's Steel Fabrications	01745 852266	Robert Howie & Sons	01560 484941
B & K Steelwork Fabrications Ltd	01773 853400	Jackson Steel Structures Ltd	01382 858439
A C Bacon Engineering Ltd	01953 850611	James Bros (Hamworthy) Ltd	01202 673815
Ballykine (Structural Engineers) Ltd	01238 562560	Joy Steel Structures (London) Ltd	0171 474 0550
Barnshaw Section Benders Ltd	01902 880848	James Killelea & Co Ltd	01706 229411
Barrett Steel Buildings Ltd	01274 682281	T A Kirkpatrick & Co Ltd	0146 1800 275
Billington Dewhurst Ltd	01942 817770	Knight & Butler Ltd	01342 832132
Billington Structures Ltd	01226 340666	The Lanarkshire Welding Co Ltd	01698 264271
Bison Structures Ltd	01666 502792	Terence McCormack Ltd	01693 62261
Bone Steel Ltd	01698 373331	Madden Steel Erectors Ltd	01236 424213
Booth Industries Ltd	01204 366333	Maldon Marine Ltd	01621 859000
Border Steelwork Structures Ltd	01228 48744	Harry Marsh (Engineers) Ltd	0191 534 6917
Bourne Steel Ltd	01202 746666	Midland Steel Structures Ltd	01203 445584
Britannia Fabrications Ltd	0116 269 6000	Mifflin Construction Ltd	01568 613311
W S Britland & Co Ltd	01304 831583	Modern Engineering (Bristol) Ltd	01454 318181
Henry Brook & Co Ltd	01484 421456	Monk Bridge Construction Co Ltd	01904 608416
Butler Building Systems Ltd	01592 652300	NEA Ltd Engineering	01429 235809
Butterley Engineering Ltd	01773 746111	Harold Newsome Ltd	0113 257 0156
Caunton Engineering Ltd	01773 531111	Norsteel Structures Ltd	01604 32461
Compass Engineering Ltd	01226 298388	Nusteel Structures Ltd	01303 268112
Conder Structures Ltd	01283 545377	Oldroyd Bros Ltd	01928 710666
Leonard Cooper Ltd	0113 270 5441	Harry Peers Steelwork Ltd	01204 528393
Cordforth Engineering Ltd	01642 769526	Phillips Structures Ltd	01432 267661
Coventry Construction Ltd	01203 464484	Pilot Engineering Co Ltd	01232 663612
Custom Metal Fabrications Ltd	0181 844 0940	Pring & St Hill Ltd	0117 966 3042
Frank H Dale Ltd	01568 612212	Qmec Ltd	01246 822228
Dew Group Ltd	0161 624 8291	John Reid & Sons (Strucsteel) Ltd	01202 483333
Dyer (Structural Steelwork) Ltd	01865 777431	William Reid Engineering Ltd	01309 672175
Elland Steel Structures Ltd	01422 380262	J Robertson Engineer	01255 672855
Emmett Fabrications Ltd	01274 597484	Robinson Construction	01332 574711
Evadx Ltd	01745 336413	Rowecord Engineering Ltd	01633 250511
Fairport Steelwork Ltd	01257 483311	Rowen Structures Ltd	01623 558558
Fisher Engineering Ltd	01365 388521	Salmor Structures Ltd	01846 699699
Francis & Lewis International Ltd	01452 616500	Severfield-Reeve Structures Ltd	01845 577896
Gibbs Engineering Ltd	01278 455253	William Sharp (Structural) Ltd	01922 27531
Gilcomston Construction Ltd	01224 630034	Shipley Fabrications Ltd	01476 63734
Glentworth Fabrications Ltd	01734 772088	Henry Smith (Constr. Engrs) Ltd	01606 592121
Glosford Metal Constructions Ltd	01905 723481	Snashall Steel Fabrications Co Ltd	01258 817722
S C Graham (Str. Steelwork) Ltd	01846 664089	South Durham Structures Ltd	01388 777350
Gregg & Patterson (Engineers) Ltd	01232 618131	Taylor & Russell Ltd	01772 782295
William Haley Engineering Ltd	01278 760591	Thircon Ltd	01845 522760
William Hare Ltd	01204 526111	Thomas Steelwork Ltd	0117 965 7441
Hasler Hawkins Ltd	01992 712011	Traditional Structures Ltd	01543 468100
M Hasson & Sons Ltd	01266 571281	Tube Engineering (Bristol) Ltd	01454 314201

Tubemasters Ltd	01904 692726
Ward Building Systems Ltd	01944 710421
Watson Steel Ltd (Bolton)	01204 699999
Watson Steel Ltd (Bristol)	0117 969 5361
Wescol Midlands	01384 480840
Wescol Structures Ltd	01422 203522
Westok Structural Services Ltd	01924 264121
WIG Engineering Ltd	01869 350200
Richard Wood Engineering Ltd	01761 413968
H Young Structures Ltd	01953 601881

Associate Members

AceCad Software Ltd	01628 822900
ADAMS Ltd	01203 694373
Albion Sections Ltd	0121 553 1877
ASD Anderson Brown	0131 459 3200
ASD Johnson	0113 254 3994
ASD Randle	0121 520 1231
ASD Yeovil	01963 62646
Austin Trumanns Steel Ltd	01902 351331
Ayrshire Metal Products Ltd	01327 300990
British Steel plc	0171 735 7654
SP&CS - Redcar	01642 404040
SP&CS - Motherwell	01698 266233
SP&CS - Scunthorpe	01724 280280
Technical - Rotherham	01709 820166
Regional Office - Leeds	0113 232 6964
Regional Office - Halesowen	0121 585 5522
Regional Office - Warrington	01925 822838
Regional Office - Belfast	01232 231821
Strip Products - Newport	01633 290022
Tubes & Pipes - Corby	01536 402121
BSD Steel Service Centres	0121 520 8844
Glasgow	0141 959 1212
Blaydon on Tyne	0191 414 2181
Stourton	0113 276 0660
Stockport	0161 483 1041
Wednesbury	0121 505 1234
Oldbury	0121 552 6812
Grantham	01476 66830
Newport	01633 270700
Bristol	01454 315314
Dartford	01322 227272
Brandon	01842 810561
Northern Ireland	01846 603604
Composite Contracting Ltd	01202 659237
Composite Contracting Ltd	01592 720005
Computer Services Consultants (UK) Ltd	0113 239 3000
Croda Mebon Ltd	01623 511000
Fundia Ltd	0121 709 0110

Hi Span Ltd	01953 603081
International Paint Ltd	0191 469 6111
W & J Leigh & Co	01204 521771
W & J Leigh & Co	0191 455 7700
T W Lench Structural Ltd	0121 559 1530
Mannesmann Demag Material Handling Ltd	01295 264555
Metsec Plc	0121 552 1541
Newton Steelstock Ltd	01963 363763
Pillar-Wedge Group Ltd	01274 737153
Precision Metal Forming Ltd	01242 527511
Press & Shear Machinery Ltd	01827 250000
Pullmax Ltd	0113 277 8330
Structural Metal Decks Ltd	01425 471088
Structural Sections Ltd	0121 558 3222
Studwelders Ltd	01291 626048

Corporate Members

The University of Leeds	0113 233 2295
Dr R J Pope Consulting Engineer	01752 263636
R Stainsby	01642 722589
Griffiths & Armour	0151 236 5656
Balfour Kilpatrick Ltd	01332 661491
Highways Agency	0171 921 4634
Joseph Murphy Str. Engineers Ltd	01253 872353
The Vinden Partnership	01204 362888